AF839682

# Autogenous Healing of Concrete in the Drinking Water Industry

Awwa
Research
Foundation

## About the Awwa Research Foundation

The Awwa Research Foundation (AwwaRF) is a member-supported, international, nonprofit organization that sponsors research to enable water utilities, public health agencies, and other professionals to provide safe and affordable drinking water to consumers.

The Foundation's mission is to advance the science of water to improve the quality of life. To achieve this mission, the Foundation sponsors studies on all aspects of drinking water, including supply and resources, treatment, monitoring and analysis, distribution, management, and health effects. Funding for research is provided primarily by subscription payments from approximately 1,000 utilities, consulting firms, and manufacturers in North America and abroad. Additional funding comes from collaborative partnerships with other national and international organizations, allowing for resources to be leveraged, expertise to be shared, and broad-based knowledge to be developed and disseminated. Government funding serves as a third source of research dollars.

From its headquarters in Denver, Colorado, the Foundation's staff directs and supports the efforts of more than 800 volunteers who serve on the board of trustees and various committees. These volunteers represent many facets of the water industry, and contribute their expertise to select and monitor research studies that benefit the entire drinking water community.

The results of research are disseminated through a number of channels, including reports, the Web site, conferences, and periodicals.

For subscribers, the Foundation serves as a cooperative program in which water suppliers unite to pool their resources. By applying Foundation research findings, these water suppliers can save substantial costs and stay on the leading edge of drinking water science and technology. Since its inception, AwwaRF has supplied the water community with more than $300 million in applied research.

More information about the Foundation and how to become a subscriber is available on the Web at **www.awwarf.org.**

# Autogenous Healing of Concrete in the Drinking Water Industry

Prepared by:
**Jeffrey Parks**, **Marc Edwards**, **Peter Vikesland**, **Matthew Fiss**, and **Abhijeet Dudi**
Virginia Polytechnic Institute and State University
Department of Civil and Environmental Engineering, 418 Durham Hall, Blacksburg, VA 24061

Sponsored by:
**Awwa Research Foundation**
6666 West Quincy Avenue, Denver, CO 80235-3098

Published by:

Awwa Research Foundation

Distributed by:

American Water Works Association

IWA Publishing

**DISCLAIMER**

This study was funded by the Awwa Research Foundation (AwwaRF). AwwaRF assumes no responsibility for the content of the research study reported in this publication or for the opinions or statements of fact expressed in the report. The mention of trade names for commercial products does not represent or imply the approval or endorsement of AwwaRF. This report is presented solely for informational purposes.

Copyright © 2008
by Awwa Research Foundation

ALL RIGHTS RESERVED.
No part of this publication may be copied, reproduced
or otherwise utilized without permission.

# CONTENTS

# TABLES

# FIGURES

# FOREWORD

The Awwa Research Foundation is a nonprofit corporation that is dedicated to the implementation of a research effort to help utilities respond to regulatory requirements and traditional high-priority concerns of the industry. The research agenda is developed through a process of consultation with subscribers and drinking water professionals. Under the umbrella of a Strategic Research Plan, the Research Advisory Council prioritizes the suggested projects based upon current and future needs, applicability, and past work; the recommendations are forwarded to the Board of Trustees for final selection. The foundation also sponsors research projects through the unsolicited proposal process; the Collaborative Research, Research Applications, and Tailored Collaboration programs; and various joint research efforts with organizations such as the U.S. Environmental Protection Agency, the U.S. Bureau of Reclamation, and the Association of California Water Agencies.

This publication is a result of one of these sponsored studies, and it is hoped that its findings will be applied in communities throughout the world. The following report serves not only as a means of communication of the results of the water industry's centralized research program but also as a tool to enlist the further support of nonmember utilities and individuals.

Projects are managed closely from their inception to the final report by the foundation's staff and large cadre of volunteers who willingly contribute their time and expertise. The foundation serves a planning and management function and awards contracts to other institutions such as water utilities, universities, and engineering firms. The funding for this research effort comes primarily from the Subscription Program, through which water utilities subscribe to the research program and make an annual payment proportionate to the volume of water they deliver and consultants and manufacturers subscribe based on their annual billings. The program offers a cost-effective and fair method for funding research in the public interest.

A broad spectrum of water supply issues is addressed by the foundation's research agenda: resources, treatment and operations, distribution and storage, water quality and analysis, toxicology, economics, and management. The ultimate purpose of the coordinated effort is to assist water suppliers in providing the highest possible quality of water economically and reliably. The true benefits are realized when the results are implemented at the utility level. The foundation's trustees are pleased to offer this publication as a contribution toward that end.

David E. Rager
Chair, Board of Trustees
Awwa Research Foundation

Robert C. Renner, P.E.
Executive Director
Awwa Research Foundation

# ACKNOWLEDGMENTS

The authors of this report wish to thank Maureen Hodgins, AwwaRF project officer, and the members of the Project Advisory Committee – Thomas Rockaway, University of Louisville; David Gress, University of New Hampshire; and Edgar Navera, City of Philadelphia Water Department.

We also appreciate the efforts of Paolo Scardina for his assistance with the SEM-EDS analyses and Robert Simonds for helping with the strength testing. Steven Abbott also provided assistance with the many water changes throughout the study.

# EXECUTIVE SUMMARY

Corrosive degradation of materials through physicochemical interaction with the environment has profound economic, social, and public health consequences. Unfortunately, little research expenditure has gone towards the investigation of processes responsible for the corrosion of concrete drinking water infrastructure. One important mode of attack on concrete involves access of corrosive water to underlying reinforcing steel through cracks, which then corrodes, expands and causes failure via spalling. Under some circumstances it is known that cracks in the concrete surface can repair themselves through reactions with constituents in the water. This phenomenon has been termed "autogenous healing" and while it is known to be important, only a few studies conducted over the past 50 years even attempt to explain the mechanism behind this 'self-sealing'. Most previous researchers are in agreement that concrete must be submerged in water for autogenous healing to occur and that cracks are closed by the precipitation of calcium carbonate and possibly calcium hydroxide. Some investigators postulated that the chemistry of the water contacting the pipe had little to do with the process. In the present study, however, water chemistry played a pivotal role in determining the extent of autogenous healing. Additionally, the presence of magnesium and silicon were critical for a strong 'glue' to form in artificially constructed cracks.

## RESEARCH OBJECTIVE

The Awwa Research Foundation (AwwaRF) funded this study to systematically investigate the phenomenon of autogenous healing. Work was performed in three distinct tasks. In Task 1, the effect of water chemistry on concrete corrosion and autogenous healing was evaluated. Attempts were made to identify precipitates responsible for autogenous healing via scanning electron microscopy and Raman spectroscopy in Task 2. In Task 3, a determination of the level of restoration was made by evaluating the strength of the 'glue' that formed on the crack surfaces, the presence or absence of carbonation on internal crack surfaces, and the ability of the 'glue' to impede chloride diffusion through concrete specimens.

Although strength restoration is an important consideration when evaluating the effectiveness of autogenous healing, the prevention of carbonation and chloride diffusion are equally as important. Carbonation is the result of attack by carbon dioxide and occurs when carbon dioxide dissolves in water, forming carbonic acid. This acid neutralizes alkali in concrete pore water resulting in the dissolution of lime ($Ca(OH)_{2(s)}$) and the precipitation of calcium carbonate. Ultimately, the pH in the pore water begins to fall and this causes corrosion of underlying steel in reinforced concrete. The prevention of chloride diffusion through concrete cracks is important for a similar reason. Chloride can attack underlying steel in reinforced concrete causing premature failure of concrete infrastructure.

## APPROACH

A new technique was developed to simulate autogenous repair of cracks in concrete structures. Concrete for this study was prepared from one-half inch maximum coarse aggregate and portland cement. Our basic approach was to cut small coupons of concrete into two halves and control the crack width through the use of 0.25 mm (0.01 in) thick polyethylene spacers. These coupons were suspended in two liter containers with water containing various combinations of calcium, magnesium, phosphorus, silicon, zinc, natural organic matter (NOM), and carbonate at pH 7, 8, 8.5, or 9.5. A total of 34 water chemistry combinations were tested in a 12 month screening study. Results of the screening study were used to select conditions for a second in-depth study utilizing 26 water chemistry combinations. Experiments were conducted in triplicate by suspending three concrete specimens in each container. The contents of each container were gently agitated using an orbital shaker operating at 25 rpm.

## RESULTS

Results from the 12 month screening study indicate that autogenous healing is most certainly controlled by the water chemistry. The formation of a precipitate that restored strength to the concrete, or 'glued' the separated halves together, only occurred consistently in waters with high levels of magnesium and silicon. In some cases the strength of the glue was quite strong. For example, an average force of 215 N was required to break apart the autogenously healed specimens in waters that had pH 9.5, 40 mg/L magnesium and 35 mg/L silicon. SEM-EDS analyses indicated the presence of magnesium and silicon on the crack surface after these higher strength specimens were sheared apart.

Results from the subsequent 7 month in-depth study provided additional insight. Visual observations (as evidenced by the passage of light through the crack) demonstrated that some degree of autogenous healing occurred in all water chemistries tested. The highest strength specimens from this study were once again those in pH 9.5 water containing high levels of magnesium and silicon. While SEM-EDS and Raman analyses show that calcium carbonate and/or calcium hydroxide are present in the precipitate formed in many of the water chemistries tested, the 'glue' in the strongest specimens always contained a magnesium silicate component.

The extent of healing that occurs can be quantified in a number of ways. First, the most beneficial type of autogenous healing involves filling of the crack with restoration of strength with a magnesium silicate solid. In other instances, the crack was filled by precipitates that did not restore strength to the separated sections of concrete, but nonetheless, could serve as a barrier to diffusion of chloride and other corrosive constituents (e.g., $CO_2$) to the underlying rebar and concrete. Formation of a physical barrier without restoring strength was shown to significantly impede transport of chloride in this study.

## CONCLUSIONS

Conclusions that can be taken from this work include:

- Autogenous healing can be controlled by water chemistry, and it can be an effective method of sealing small cracks before they develop into larger cracks that can cause failure.
- Even though a high strength seal can be developed between the cracks, the formation of a precipitate on concrete surfaces at pH 9.5 when high levels of magnesium and silicon are present does not protect concrete from uniform corrosion. Thus, protection from general corrosion of concrete and autogenous repair are not necessarily inter-related. It is possible to have a water with high rates of general concrete attack with a high likelihood of autogenous repair of cracks.

- Carbonation of internal concrete surfaces was not detected when cracks were sealed via autogenous healing.
- Chloride diffusion rates do not return to levels seen in virgin concrete when cracks are autogenously healed. However, water permeability and chloride diffusion are impeded by filling of cracks, regardless of strength attained.

## RECOMMENDATIONS

Future research should test other hydraulic conditions and water chemistries that might enhance or impede autogenous repair. Given that autogenous repair occurred more readily at higher pH values, it is anticipated that high velocity low pH water might decrease the rate of crack repair.

Strength data for shorter exposure periods could provide additional insight into the autogenous healing process. White precipitates on crack surfaces were visible after as little as 6 weeks of exposure in some situations. A follow-up study in which strength tests could be conducted on sacrificial coupons at 30 day intervals would be invaluable. One caveat, however, is that a high level of variability was observed in these tests and therefore, results from single sacrificial coupons may not be indicative of the degree of autogenous healing of other test coupons subjected to the same water quality parameters.

The authors believe various factors attributed to the high degree of variability exhibited by replicate coupons. A major factor appeared to be the heterogeneity of the concrete specimens. Coupons were a fairly small size and the presence or absence of large aggregates could affect a coupon's response to various water quality parameters. Future research should perhaps eliminate the large aggregate fraction and utilize only cement, sand, and water to cast concrete coupons.

While similar trends were observed in both series of tests, the four-fold higher strength gain observed in the 12 month study as compared to the subsequent 7 month study is a concern. While varying solution chemistries may account for some of this inconsistency, the authors feel that the test conditions greatly affected this outcome. Large stagnation intervals and limited pH control in the 12 month study would be expected to be conducive for the occurrence of autogenous healing. That is, during periods of stagnation the pore water pH could increase dramatically which could then lead to higher levels of precipitates responsible for the autogenous healing phenomenon. Future research, as stated above, should make a distinction between stationary and flowing water.

Concrete age could be another factor that might enhance or impede autogenous repair. Coupons used in this study were 3-5 years old. Older concrete, such as that found in many distribution systems, may behave differently.

The use of the Langelier Index (LI) to determine whether concrete corrosion will occur in distribution systems is widely accepted although it is not always reliable. In this study we have observed that LI cannot indicate when appreciable autogenous healing of cracks in concrete will occur with restoration of strength. Obviously, other factors such as magnesium and silicon concentrations in the treated water play an important role and must be explicitly considered if an index were to be developed that describes autogenous repair.

For utilities with low amounts of magnesium and silicon in their treated water, we speculate that high rates of concrete degradation could occur due to the absence of autogenous healing. Small cracks that tend to grow over time without autogenous healing will eventually increase the likelihood of failures. In such cases a liner or seal coat might be increasingly important in preventing unreasonable rates of failure. While it may be possible to promote healing by manipulating the environment (water chemistry) or the concrete composition, the interplay between these two factors needs to be further evaluated.

# CHAPTER 1: INTRODUCTION AND BACKGROUND

Corrosion, the degradation of materials through physicochemical interaction with the environment, has profound economic, social, and public health consequences. In the United States it is estimated that the annual cost of corrosion ranges from $200 to $271 billion, or about 3.1% of the nation's gross domestic product (US Department of Transportation 2002). As a result of these societal costs, a significant research effort has characterized atmospheric corrosion processes, the corrosion of metal reinforced concrete bridges and roads, and corrosion in the defense industry. Unfortunately, comparatively little research expenditure has gone towards the investigation of processes responsible for the corrosion of concrete drinking water infrastructure. An improved understanding of the mechanisms involved and the ultimate mitigation of concrete corrosion are of significant importance in light of its social and economic costs (McNeill and Edwards 2003; Jackson, Pitt, and Scabo 1992; Dingus, Haven, and Austin 2002).

## AUTOGENOUS HEALING

Though poorly studied, it is typically believed that small cracks in concrete pipe and concrete liners allow access of corrosive water to reinforcing metal, which then corrodes, expands and causes pipe failure via spalling or attack on reinforcing metal (Dingus, Haven, and Austin 2002; McNeill and Edwards 2003). Under some circumstances, the cracks that develop in concrete can repair themselves through reactions with constituents in the water (Clear 1985, Neville, 2002, Ramm and Biscoping 1998, Hearn 1998, Edvarsen 1999, Aldea et al. 2000). This phenomenon has been termed "autogenous healing" and several studies have been conducted over the past 50 years that have attempted to explain the mechanism behind this 'self-sealing' (Table 1.1). Most previous researchers are in agreement that concrete must be submerged in water for autogenous healing to occur and that cracks are closed by the precipitation of calcium carbonate and possibly calcium hydroxide. At least one researcher has proposed that the migration of loose particles (mechanical blocking) is partially responsible (Clear 1985), while others assert that the precipitation of calcite is almost solely responsible (Edvarsen 1999).

## CHARACTERISTICS OF CONCRETE

Concrete is a mixture of two components: aggregate and binder. The aggregate provides the basic structure of the material and is typically either sand or gravel. The binder, or mortar, consists of hydrated cement and is responsible for binding the aggregate together and thus provides the cohesive properties of the material. Dry powdered cement is produced by calcinating calcareous (calcium containing) and argillaceous (clay-containing) rocks with either silica or alumina (AwwaRF and DVGW-TZW 1996). The cement produced by this process contains a number of silicates and oxides of calcium, aluminum, and iron. The portland cement commonly employed in water treatment and distribution primarily consists of tricalcium silicate ($Ca_3SiO_5$), dicalcium silicate ($Ca_2SiO_4$), calcium aluminate ($Ca_3Al_2O_6$), iron calcium aluminate ($Ca_4Fe_2Al_2O_{10}$), and gypsum ($CaSO_4 \cdot 2H_2O$).

**Table 1.1**
**Previous studies pertaining to autogenous healing of concrete**

| Study | Crack Description | Environment | Concrete Age | Test Duration | Final Condition | Author's Conclusions |
|---|---|---|---|---|---|---|
| Lauer and Slate, 1956 | Not reported (specimens broken and held together with rubber band) | Air with RH 95% or submerged in water (composition not reported) | up to 90 days | 6 mo | 25% strength regain by submerging in water for 89 days; storage in humid air resulted in strengths as much as 85% lower than for submerged specimens. | Specimens stored in air developed less strength more slowly. Calcium carbonate and calcium hydroxide ppt responsible for healing. |
| Wagner, 1974 | Rosette pattern with max crack width of 0.08 mm | Stagnant water, pH 6.8, 21 mg/L Ca, alkalinity 52 mg/L | unknown; commercially available items tested | 60 d | Sealed with white ppt. identified as calcium carbonate | Autogenous healing occurs in both flowing and non-flowing conditions, but occurs faster in stagnant water. |
| | Single with max width of 0.33 mm | Stagnant water, pH 6.8, 21 mg/L Ca, alkalinity 52 mg/L | | 30 d | Partially filled with white ppt. identified as calcium carbonate | |
| | Numerous with max width of 0.10 mm | Flowing water, pH 6.8, 21 mg/L Ca, alkalinity 52 mg/L | | 365 d | Width reduced to 0.05 mm after 180 d with unidentified brown ppt; 0.10 mm crack healed below surface and all smaller cracks healed in 1 yr | |

(continued)

**Table 1.1 (continued)**

| Study | Crack Description | Environment | Concrete Age | Test Duration | Final Condition | Author's Conclusions |
|---|---|---|---|---|---|---|
| Clear, 1985 | 0.20 mm and 0.15 mm crack widths; concrete thk 0.445 - 0.460 m | Concrete reservoir; pH 8.13, 112 mg/L Ca; constant head of 2.85 m or 3.03 m | 2 - 3 d | 18 d | pH increase to 12.77 and Ca increase to 505 mg/L after 36 hours; thereafter pH and Ca levels decreased; white ppt formed on outside surface | Healing occurs by combination of mechanical blocking with small particles and ppt of calcium carbonate. |
| | 0.30 mm max width; 67 mm long x 150 mm deep | Hard water; pH 7 - 7.9, 56 - 76 mg/L Ca; constant head of 3.37 m | 30 - 56 d | 7 d | pH increase to 12.4 through small cracks; high levels of Ca through all cracks | Ppt of calcium carbonate is not predominant mechanism which reduces flow during early stages of flow. Healing mechanism is a combination of mechanical blocking and ppt of calcium carbonate. |
| | 0.30 mm max width; 67 mm long x 150 mm deep | Soft water; pH 7.5 - 8, 2 - 36 mg/L Ca; constant head of 3.37 m | 42 - 62 d | | pH increase to 12.2 through small cracks; high levels of Ca through all cracks | |
| | 0.29 mm max width; 67 mm long x 150 mm deep | Distilled water; pH 6.5, 0 mg/L Ca; constant head of 1.0 m | 35 - 56 d | | Layer of calcium carbonate visible on crack inner surfaces; quality of water through crack not measured | |

(continued)

**Table 1.1 (continued)**

| Study | Crack Description | Environment | Concrete Age | Test Duration | Final Condition | Author's Conclusions |
|---|---|---|---|---|---|---|
| Munday et al., 1974 | Formed by compressive loading device; qty, width, and length not reported | 3 conditions tested: air with RH = 55%; air with RH = 92-95%; water at 20 C (composition not reported) | 7 - 240 d | 240 d | High humidity condition had best regain of strength | Moist environment can result in full strength recovery |
| Hearn and Morley, 1997 | Not reported | Flowing water, pH 7.5, 30 mg/L Ca | paste 20 h; mortar 18 mo; concrete 26 yr | 7 d | Permeability decreased to nearly constant value in 4 d for 26 yr concrete | Migration of loose particles was not responsible for SSE (self sealing effect); dissolution and deposition is primary mechanism for SSE |
| Aldea et al., 2000 | 0.014 - 0.350 mm; (specimen: 100 mm diameter; 25 or 50 mm deep) | Stagnant water (composition not reported) | 28 d | 100 d | Final permeability of cracked concrete (up to 0.20 mm width) is approx same as uncracked. Full strength regain not observed. | Material responsible for decrease in permeability not as strong as original uncracked material. |

(continued)

**Table 1.1 (continued)**

| Study | Crack Description | Environment | Concrete Age | Test Duration | Final Condition | Author's Conclusions |
|---|---|---|---|---|---|---|
| Ramm and Biscoping, 1998 | 0.1, 0.2, 0.3, 0.4 mm width / 30 or 60 mm deep | Flowing water, deionized, pH 7.0; constant head of 2.5 m or 12 m | not reported | 2 yr | Water emerging as individual drops at end of test | Complete healing not observed under any of these test conditions. |
| | | Flowing water, boric acid-treated, pH 6.1; constant head of 2.5 m or 12 m | | | | |
| | | Flowing water, boric acid-treated, pH 5.2; constant head of 2.5 m or 12 m | | | Permeability increased throughout test | |
| Edvardsen, 1999 | 0.1, 0.2, 0.3 mm width / 200, 300, 400 mm deep | Flowing water (soft), 28 mg/L Ca; head range 2.5 - 20 m; or (hard), 192 mg/L Ca; head range 2.5 - 20 m | | 5 - 20 wks | 50% of specimens with crack width 0.20 mm and head 2.5 m healed completely in 7 wks. 25% of specimens with crack width 0.20 mm and head 10 m healed completely in 7 wks. | Calcite formation in crack almost solely responsible for healing. Concrete composition and water hardness have no effect. |

When portland cement is wetted, these phases undergo a series of hydrolysis, hydration, and ultimately precipitation reactions that produce a hardened material. This process, which can take many years, results in the formation of calcium hydroxide ($Ca(OH)_2$) and poorly crystalline calcium silicate hydrate (C-S-H) gel. These phases are the 'glue' that binds the aggregate together and it is their stability that dictates the corrosion resistance of concrete (Deng et al. 2002).

During the hardening period, the interstitial water within the hydrated cement matrix is in equilibrium with the precipitating phases and contains high concentrations of calcium, magnesium, sodium, and hydroxide. The pH of the interstitial water is typically between 12 and 13.5 and in carbonate containing solutions, calcium carbonate ($CaCO_3(s)$) precipitates readily both at the concrete-water interface and within the micro-pores that extend from the concrete interior to the interface (Figure 1.1). In general, this precipitated $CaCO_3(s)$ (typically presumed to be the mineral calcite, although the $CaCO_3(s)$ polymorphs aragonite or vaterite are also possible) has been assumed to protect the underlying concrete from degradation (Douglas, Merrill, and Catlin 1996) and as discussed previously, is assumed to be responsible for the phenomenon of autogenous healing.

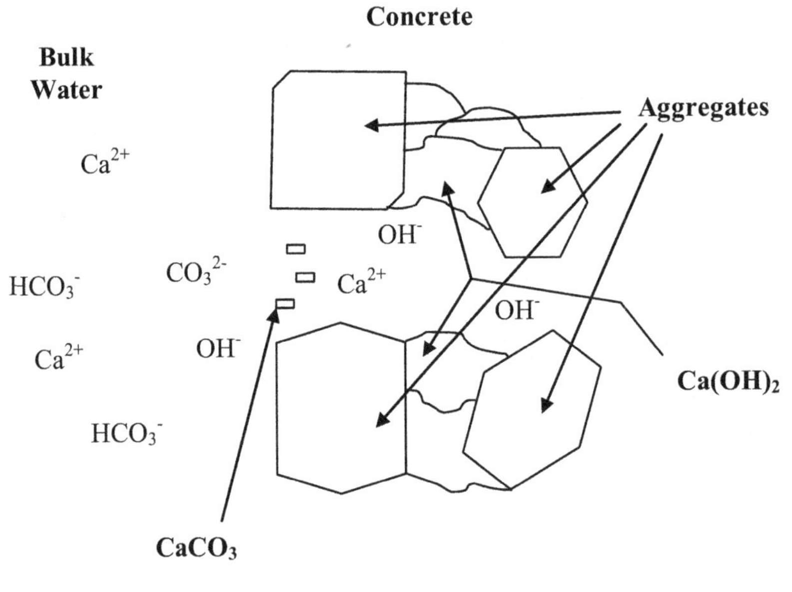


**Figure 1.1 Conceptualization of a concrete micro-pore that is actively corroding**

## THE LANGELIER INDEX

As a result of its heterogeneous composition, the corrosion of concrete is an exceptionally complex process. However, as noted previously, it is commonly believed that the presence of a $CaCO_3(s)$ layer at the concrete-water interface is protective. Accordingly, one of the key factors used to evaluate the potential for concrete degradation is the Langelier Index (LI; AwwaRF and DVGW-TZW 1996; McNeill and Edwards 2003). LI is a measure of the degree to which water is able to precipitate calcium carbonate and can be calculated from Equations 1.1 and 1.2.

$$LI = pH_a - pH_s \tag{1.1}$$

$$pH_s = pK_s - \log\,[Ca^{2+}] - \log\,[HCO_3^-] - \log \gamma_{Ca^{2+}} - \log \gamma_{HCO_3^-} \tag{1.2}$$

where $pH_a$ = measured bulk pH

$pH_s$ = pH at which the solution is saturated with $CaCO_3(s)$

$pK_s$ = the solubility constant for calcite at a given temperature.

$\gamma_{Ca^{2+}}$ = activity coefficient of calcium

$\gamma_{HCO_3^-}$ = activity coefficient of bicarbonate

Waters with positive LI values are typically deemed non-aggressive towards concrete (because of their tendency to precipitate $CaCO_{3(s)}$), while those with negative LI values are thought to be aggressive (because of their tendency to dissolve $CaCO_3(s)$; Douglas, Merrill, and Catlin 1996).

Unfortunately, the use of Langelier indices to predict concrete degradation has met only limited success. Although waters that are highly undersaturated with $CaCO_3(s)$ (i.e., negative LI values) are often highly aggressive towards concrete, it is possible to find some waters with positive LI values that can corrode concrete -- in fact a recent literature review suggests that Langelier indices do not always correlate with much of the practical concrete corrosion data that has been gathered in the field (McNeill and Edwards 2003). This result can be expected since the use of LI values dramatically oversimplifies the concrete corrosion process. In particular, it does not account for the extreme stratification in pH that can exist between 'bulk water' and the concrete pore water. Concrete corrosion can initiate deep within concrete pores, and as a result the measurements of bulk pH may not reflect the pH of the region where corrosion is occurring. It is well established that concrete corrosion processes occurring at the surface may be quite different from those that occur within the concrete matrix. Another significant drawback in the use of Langelier indices is that they do not consider the potential effects of other ions such as sulfate, chloride, and magnesium on corrosion. Changes in the concentration of each of these ions has the capability to affect concrete corrosion rates (McNeill and Edwards 2003), presumably by formation of other solids that can protect the underlying pipe.

## EFFECTS OF MAGNESIUM AND SILICON

The role of magnesium in concrete corrosion has not been well addressed in the literature. It has been hypothesized that the $Mg^{2+}$ ion could readily react with $OH^-$ ions present in the concrete pore water to form the mineral brucite ($Mg(OH)_2(s)$) (Santhanam, Cohen, and Olek 2002). This, in turn, maintains a lower pH in pore water that could enhance $Ca(OH)_2$ and C-S-H dissolution, thereby enhancing concrete degradation. Additionally, Kvech and Edwards (2002) discovered that magnesium-aluminum oxy-hydroxides were highly insoluble near pH 9.5 during water softening processes. Since the pH near the surface and within pores of concrete is often higher than that of bulk water, it is deemed likely that similar compounds could form on concrete via reaction between aluminum (from the dissolving cement) and magnesium (from the water). Magnesium silicates are also known to form during water softening processes, and to precipitate in pipelines in some cases (Gunnlaugsson and Einarsson 1989; Kristmannsdottir, Olafsson, and Thorhallsson 1989), and therefore could be expected to form on dissolving cement as well (Kristiansen 1977, Price et al. 1997).

Recent unpublished work in Europe has determined that aqueous silicate (Slats 2002) tends to decrease aluminum leaching to potable water from the cement. The implication is that aluminum silicates, or perhaps aluminum magnesium silicates (Kvech and Edwards 2002) are forming at the concrete surfaces, and that the silicate can be protective to concrete. Literature reviews (AWWA 2001) have also suggested that phosphorus or zinc inhibitors can be protective to concrete. However, there is nothing definitive in the literature proving cause, effect or mechanism of such protection, if it is indeed significant.

## POTENTIAL RELEVANCE

The drinking water industry has been slow to study the phenomenon of self-healing concrete. Although first discovered in 1836 by the French Academy of Science, few studies have examined the potential importance of autogenous repair to drinking water treatment and transmission (Table 1.1). The rate at which concrete corrodes controls the longevity of water treatment plants, cement water mains and ductile iron pipe lined with cement. In addition, autogenous healing may lead to lower chloride diffusion (from the transported water through the concrete infrastructure), which in turn would provide corrosion protection for underlying reinforcing steel. Accordingly, a thorough understanding of the processes responsible for concrete corrosion is of fundamental concern to the water treatment industry.

Our objective in this study was to systematically investigate the phenomenon of autogenous healing. By carefully examining the effects of bulk water chemistry on concrete corrosion and autogenous repair, we intend to develop groundwork that could provide better guidance for the water industry to predict the aggressiveness of a given water toward their concrete infrastructure.

## ORGANIZATION OF THE REPORT

This report is organized as follows:

- Chapter 1 introduces the idea of autogenous healing in concrete and summarizes the literature regarding past researchers' attempts at explaining its cause.
- Chapter 2 describes the methods and materials used in this study in which bench scale tests are conducted that will allow us to define water chemistry parameters that may be relevant for autogenous healing to occur.
- Chapter 3 presents results from a 12 month 'screening' study. This study was designed to test a wide range of water quality parameters.
- Chapter 4 presents results from a follow-up 7 month 'in-depth' study.
- Chapter 5 describes the results from mechanical properties tests. These include the shear strength of the autogenous 'glue' which formed during the bench scale tests, as well as tests to quantify the ability of the autogenous 'glue' to impede chloride diffusion.
- Chapter 6 describes our attempts to identify the composition of the autogenous 'glue'.
- Chapter 7 summarizes the conclusions of the project and provides our recommendations to utilities.

# CHAPTER 2: METHODS AND MATERIALS

A new technique was developed to simulate autogenous repair of cracks in concrete structures. Our basic approach was to cut small coupons of concrete into two halves and control the crack width through the use of 0.25 mm (0.01 in) thick polyethylene spacers. Previous researchers have primarily used a compression device to develop cracks of various widths (e.g. Munday, Sangha, and Dhir 1974; Clear 1985; Edvarsen 1999). While our method may not be analogous to cracks that occur in practice, we feel that we have developed a 'worst case' scenario by eliminating the concept that continued hydration may be responsible for healing. It also causes complete separation and loss of structural strength between the two sections of concrete, in which case any restoration of strength or filling of gaps can be unambiguously attributed to autogenous repair.

## CONCRETE PREPARATION

Concrete for this study was prepared for a prior study in 2000 from one-half inch maximum coarse aggregate and portland cement. Ingredient properties and characteristics are listed in Table 2.1. Two inch (51 mm) cubes were cast and cured in saturated lime water at 21 ± 1 °C (70 ± 2 °F) for seven days. At the time of this study the concrete had aged three to five years.

**Table 2.1**
**Concrete constituent characteristics**

| *Aggregate* | | | *Cement* | | *Mixture* | |
|---|---|---|---|---|---|---|
| **% Passing Size** | **Coarse** | **Fine** | **Analyte** | **Weight %** | **Ingredient** | **Weight (kg)** |
| 3/4 inch | 100 | 100 | $SiO_2$ | 21.25 | Cement | 269 |
| 1/2 inch | 100 | 100 | $Al_2O_3$ | 4.49 | Water | 113 |
| 3/8 inch | 50 | 100 | $Fe_2O_3$ | 3.04 | Coarse Aggregate | 686 |
| #4 | 8 | 99 | CaO | 63.51 | Fine Aggregate | 594 |
| #8 | 2 | 90 | MgO | 2.48 | | |
| #16 | - | 78 | $SO_3$ | 2.47 | | |
| #30 | - | 46 | $Na_2O$ | 0.17 | W/C ratio | 0.42 |
| #50 | - | 17 | $K_2O$ | 0.82 | | |
| #100 | - | 2 | $TiO_2$ | 0.21 | | |
| #200 | - | 0.5 | $P_2O_5$ | 0.11 | | |
| | | | $Mn_2O_3$ | 0.06 | | |
| | | | SrO | 0.14 | | |

## AUTOGENOUS HEALING TESTS

Tests were conducted by precisely cutting a 38 mm × 13 mm × 6 mm (1-1/2" × ½" × ¼") concrete coupon, weighing the coupon, and then matching surfaces that were previously joined. In this manner, aggregate lines up with aggregate and mortar with mortar (Figure 2.1). These tests utilized a gap between the matched halves of 0.25 mm (0.01 in), simulating a hypothetical crack of the same width. These coupons were suspended in two liter containers with water containing various combinations of calcium, magnesium, phosphorus, silicon, zinc, natural organic matter (NOM), and carbonate at pH 7, 8, 8.5, or 9.5 (see Table 2.2 for complete range of values tested). The source of NOM was a Lake Pleasant fulvic acid. Calcium, magnesium, phosphorus, silicon, zinc, sulfate, and carbonate stock solutions were prepared from $CaCl_2 \cdot 2H_2O$, $MgCl_2 \cdot 6H_2O$, $Na_2HPO_4 \cdot H_2O$, $Na_2SiO_3$, $ZnCl_2$, $Na_2SO_4$, and $NaHCO_3$. A total of 34 water chemistry combinations were tested in a 12 month screening study. Results of the screening study were used to select conditions for a second in-depth study utilizing 26 water chemistry combinations. Experiments were conducted in triplicate by suspending three concrete specimens in each container (Figure 2.2). The contents of each container were gently agitated using an orbital shaker operating at 25 rpm.

**Figure 2.1 Close-up of concrete test coupon. Note matching aggregate in top and bottom halves.**

**Table 2.2**
**Range of water chemistries tested**

| Parameter | Values tested | Unit |
|---|---|---|
| pH | 7, 8, 8.5, 9.5 | - |
| Calcium | 0, 20, 45, 70 | mg/L as Ca |
| Silicon | 0, 2.5, 20, 25, 35 | mg/L as Si |
| Magnesium | 0, 10, 25, 40 | mg/L as Mg |
| Phosphorus | 0, 1, 3 | mg/L as P |
| Zinc | 0, 0.5, 1 | mg/L as Zn |
| NOM | 0, 5 | mg/L as C |
| Carbonate | 0.1, 0.68, 4, 10 | mM (10, 68, 400, and 1000 mg/L as $CaCO_3$) |
| Sulfate | 0.15 (all tests) | mM (14.4 mg/L as $SO_4^{2-}$) |

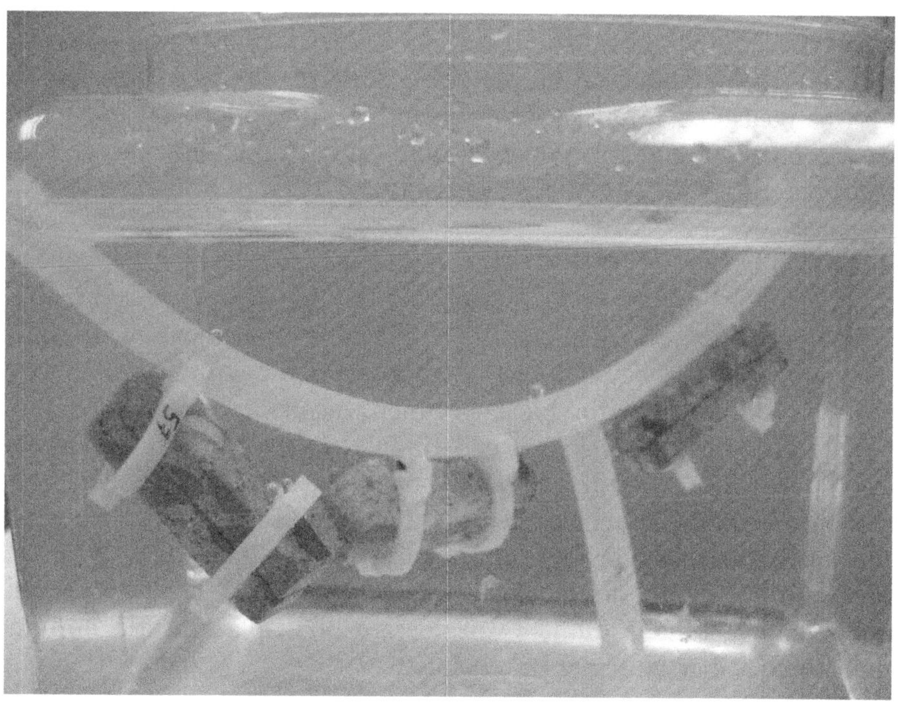

**Figure 2.2 Picture of concrete coupons in experimental apparatus**

The water in each container was changed 7 - 8 times throughout each study by dumping each two liter container and refilling with fresh water having the same chemical make-up as the initial water (Table 3.1, Table 4.1, timeline in Appendix Table A1). Initially, pH was monitored and adjusted daily (to within ± 0.2 pH units) with a Corning Model 313 with an Accumet Ag/AgCl electrode, but as the test progressed fewer adjustments were required so monitoring was performed less frequently. Periodic samples were taken for metals analysis, anion analysis, and total organic carbon (TOC) analysis. Metals (Al, Ca, Fe, Mg, P, Si, and Zn) were analyzed by inductively coupled plasma emission spectrometry (ICP-ES) in accordance with Standard Method 3120-B during the initial screening study and by inductively coupled plasma mass spectrometry (ICP-MS) in accordance with Standard Method 3125-B during the latter 7 month in-depth study (APHA 1998). Anion concentrations (chloride, sulfate, nitrate, and nitrite) were analyzed by ion chromatography per Standard Method 4110 using a Dionex DX 120 ion chromatograph (APHA 1998). TOC was measured with a Sievers 800 Portable Total Organic Carbon Analyzer. Visual observations of each specimen were also recorded periodically.

## CHLORIDE DIFFUSION

At the end of the seven month in-depth study no visible light was observed through the cracks of several specimens. A portion of these were selected and tested for their ability to inhibit chloride diffusion. A 4" (100 mm) long, ½" (13 mm) PVC tube was affixed to the concrete specimen with silicone sealant such that a portion of the crack was positioned below the tube (Figure 2.3). After the sealant had cured, a 5 M NaCl solution was poured into the tube. A 100 mL glass beaker containing 60 mL deionized water was placed beneath the specimen and conductance was measured in this solution using a YSI Model 32 Conductance Meter. The conductance was related to chloride concentration using a calibration curve. In most cases, tests were terminated after only a few minutes due to the rapid rate of diffusion.

## STRENGTH TESTING

At the conclusion of each study the concrete specimens were removed from the containers and subjected to a strength test to determine the degree to which autogenous healing was successful in restoring strength. Each specimen was placed onto a MTS Tytron 250 strength testing device fitted with a 250 N load cell (Figure 2.4). The instrument was operated in a combination displacement control and load control. After a specimen was placed in the instrument the ram was moved in displacement control until the resistant force reached 20 N. At this point the instrument switched to load control, where the load was ramped up until the specimen broke (in this mode, displacement was constant until the specimen failed). After the in-depth seven month test the instrument was operated in displacement control mode only due to some instrument issues that could not be remedied.

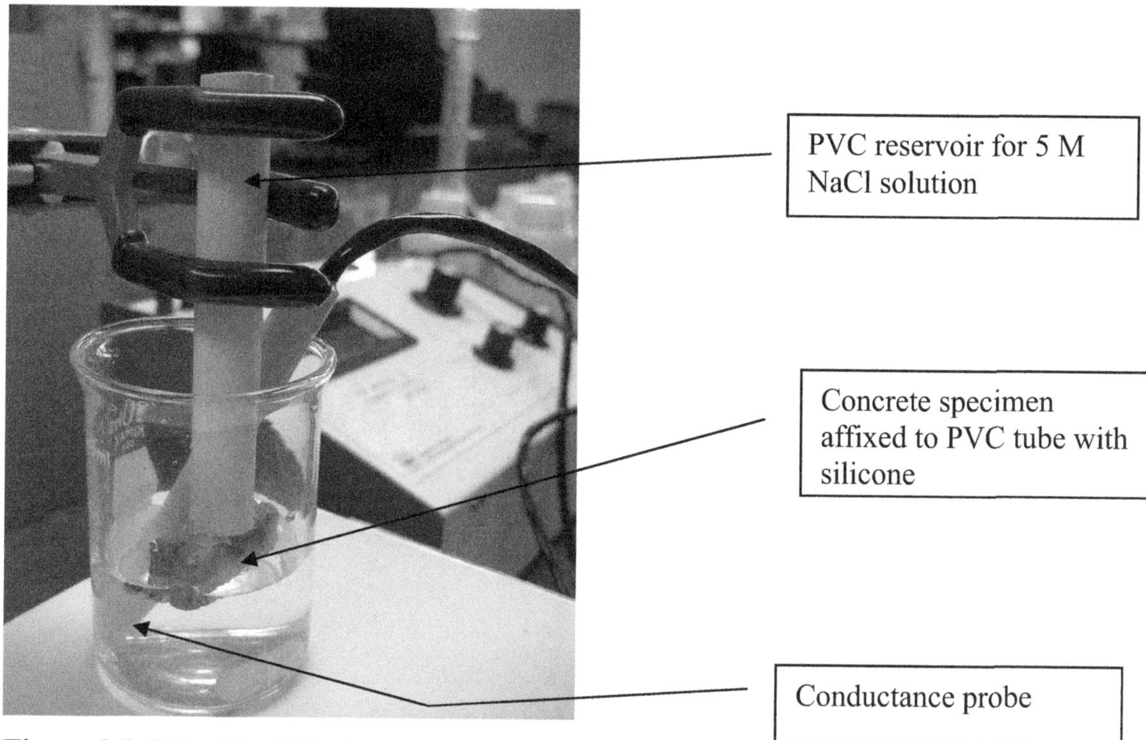


**Figure 2.3 Chloride diffusion test set-up.**

## CARBONATION TESTING

After strength testing, several specimens were also evaluated for carbonation. Carbonation is the result of attack by carbon dioxide and occurs when carbon dioxide dissolves in water, forming carbonic acid. This acid neutralizes alkali in concrete pore water resulting in the dissolution of lime ($Ca(OH)_{2(s)}$) and the precipitation of calcium carbonate. Ultimately, the pH in the pore water begins to fall. This lower pH pore water ultimately causes corrosion of underlying steel in reinforced concrete. The presence of carbonation can be visually determined by spraying a phenolphthalein indicator onto the concrete surface. A pink color indicates the surface is still alkaline and therefore not carbonated. The indicator solution is made by dissolving 1 g of phenolphthalein in 100 mL of methanol/water (50:50 v/v mix) (Broomfield 2000).

## SEM

Surfaces of some concrete specimens were analyzed at 20 kilovolts with an IXRF energy dispersive spectrometer (EDS) that was connected to a Leo 1550 field emission scanning electron microscope (FE-SEM). All samples were plated with gold and palladium at a thickness of 20 nm prior to surface analysis. Sample sites for EDS analysis were chosen at random based upon inspection with the SEM.

**Figure 2.4 Concrete specimen in strength test apparatus**

## RAMAN SPECTROSCOPY

Raman spectroscopy was also used to analyze selected specimens. Raman spectroscopy measures bond vibrations when molecules absorb electromagnetic radiation allowing identification of materials present. From this information, it is possible to determine the make-up of the 'glue' that is responsible for autogenous healing and to track the progression of crack sealing over time.

Samples were selected for Raman analysis based on visible observations of the formation of a precipitate at the crack surface. Sample spectra were obtained using a 2 second scan time repeated 30× with a 514.5 nm red laser and an objective lens with 10× magnification. The Raman instrument was calibrated prior to use by analysis of a reference slide containing pure silicon. Initially, readings were taken directly on the crack between each coupon set prior to strength testing. Following strength testing, analyses were performed on isolated constituents that appeared unique on the glue-containing face of each coupon. Specific sample locations were selected that contained different colors in rough proportion to the overall makeup of the sample coupon. It was determined that neither method allowed accurate analysis because the microscope was unable to analyze the precipitate separately from the concrete coupon (at either 10× or higher magnification). Minerals present in the concrete coupons, such as quartz, resulted in extremely high intensity spectral readings, thereby making in-situ determination of glue composition infeasible.

In order to analyze the glue, precipitates were scraped from each coupon and placed onto a glass slide. This additional step was necessary to ensure that spectra were indicative of the glue rather than the concrete coupon composition. The spectral peak location and relative peak heights were compared to spectra in Raman spectral databases (California Institute of

Technology 2005, Universita di Siena 2005). All Raman readings were performed in a completely dark room in order to reduce interference from visible light.

# CHAPTER 3: SCREENING STUDY RESULTS

A 12 month initial screening study was used to evaluate the effect of 34 different water chemistries on the autogenous healing process (Table 3.1). Water quality parameters were selected based on past studies and their perceived relevance to the autogenous healing phenomenon. Values of parameters were selected to cover the range typically present in natural waters.

## VISUAL OBSERVATIONS

Concrete specimens were evaluated visually throughout the duration of the test. As early as Day 42 it was apparent that 9 of the 34 containers had specimens with a white precipitate that had formed and filled in the crack. These included:

- pH 8.5, 25 mg/L Si and 20 mg/L Ca
- pH 8.5, 25 mg/L Si, 20 mg/L Ca, and 0.5 mg/L Zn
- pH 9.5, 25 mg/L Si and 25 mg/L Mg
- pH 9.5, 25 mg/L Si, 25 mg/L Mg, and 0.5 mg/L Zn
- pH 9.5, 25 mg/L Si, 25 mg/L Mg, and 20 mg/L Ca
- pH 9.5, 25 mg/L Si, 25 mg/L Mg, 0.5 mg/L Zn, and 1 mg/L P
- pH 9.5, 25 mg/L Si, 25 mg/L Mg, 20 mg/L Ca, and 1 mg/L P
- pH 9.5, 25 mg/L Si, 25 mg/L Mg, 20 mg/L Ca, and 0.5 mg/L Zn
- pH 9.5, 25 mg/L Si, 25 mg/L Mg, 20 mg/L Ca, 0.5 mg/L Zn, and 1 mg/L P

This accounted for 9 of 22 waters containing 25 mg/L silicon, and 9 of 19 waters containing 25 mg/L silicon at pH 8.5 or 9.5. At pH 7 none of the 3 waters containing 25 mg/L silicon had a visible precipitate. It therefore was clear that the presence of silicon in the water at higher pH was one key to repair, given that all 12 of the waters without 25 mg/L silicon did not have a precipitate filling the gap. This includes 6 waters with high levels of calcium, which were initially expected to most quickly fill the gap due to the presumed importance of $CaCO_3$.

## STRENGTH TESTING

After 12 months, all paired specimens were removed from their respective containers and were not allowed to dry until they were tested for strength restoration. Six of the 104 specimens became separated during the mild handling required before they could be mounted in the testing device, indicating that they had very low or no strength gain. The strongest set of specimens healed in water at pH 9.5 with 25 mg/L Si, 25 mg/L Mg, 20 mg/L Ca, and 1 mg/L P and required an average of 215 N (± 31 N) to break them apart (Table 3.2). The actual average is higher since one specimen did not break even with the application of a 250 N force. Concrete specimens in nine other waters required an average force greater than 100 N.

**Table 3.1**

**Composition of test waters for screening study**

| pH | Initial Concentration (mg/L) | | | | |
|---|---|---|---|---|---|
| | Ca | Mg | Si | P | Zn |
| 7 | - | - | 2.5 | - | - |
| 7 | 20 | - | 2.5 | - | - |
| 7 | - | - | 25 | - | - |
| 7 | 20 | - | 25 | - | - |
| 7 | 20 | - | 25 | 1 | - |
| 8.5 | - | - | 2.5 | - | - |
| 8.5 | 20 | - | 2.5 | - | - |
| 8.5 | - | - | 25 | - | - |
| 8.5 | 20 | - | 25 | - | - |
| 8.5 | 20 | - | 25 | - | 0.5 |
| 9.5 | - | - | 2.5 | - | - |
| 9.5 | 20 | - | 2.5 | - | - |
| 9.5 | 20 | - | 2.5 | 1 | - |
| 9.5 | 20 | - | 2.5 | - | 0.5 |
| 9.5 | 20 | - | 2.5 | 1 | 0.5 |
| 9.5 | - | - | 2.5 | 1 | - |
| 9.5 | - | - | 25 | - | - |
| 9.5 | 20 | - | 25 | - | - |
| 9.5 | 20 | 25 | 25 | - | - |
| 9.5 | 20 | - | 25 | 1 | - |
| 9.5 | 20 | 25 | 25 | 1 | - |
| 9.5 | 20 | - | 25 | - | 0.5 |
| 9.5 | 20 | 25 | 25 | - | 0.5 |
| 9.5 | 20 | - | 25 | 1 | 0.5 |
| 9.5 | 20 | 25 | 25 | 1 | 0.5 |
| 9.5 | - | 25 | 25 | - | - |
| 9.5 | - | - | 25 | 1 | - |
| 9.5 | - | 25 | 25 | 1 | - |
| 9.5 | - | - | 25 | - | 0.5 |
| 9.5 | - | 25 | 25 | - | 0.5 |
| 9.5 | - | - | 25 | 1 | 0.5 |
| 9.5 | - | 25 | 25 | 1 | 0.5 |
| 9.5 | - | - | 2.5 | - | 0.5 |
| 9.5 | - | - | 2.5 | 1 | 0.5 |

*Ca = calcium; Mg = magnesium; Si = silicon; P = phosphorus; Zn = zinc*

*0.68 mM* $HCO_3^-$ *added to all tests*

*0.15 mM* $SO_4^{2-}$ *added to all tests*

**Table 3.2**

**Strength testing results for screening study (sorted by strength within each pH block)**

| pH | Initial Concentration (mg/L) | | | | | Peak Load (N) before failure Specimen # | | | |
|---|---|---|---|---|---|---|---|---|---|
| | Ca | Mg | Si | P | Zn | 1 | 2 | 3 | Avg. |
| 7 | 20 | - | 25 | - | - | 85 | 35 | <20* | 43 |
| 7 | - | - | 25 | - | - | 42 | 48 | <20 | 33 |
| 7 | 20 | - | 25 | 1 | - | <20 | <20 | 63 | 28 |
| 7 | - | - | 2.5 | - | - | <20 | <20 | <20 | <20 |
| 7 | 20 | - | 2.5 | - | - | <20 | <20 | <20 | <20 |
| 8.5 | 20 | - | 25 | - | 0.5 | 75 | 109 | 157 | 114 |
| 8.5 | 20 | - | 25 | - | - | 40 | <20 | <20 | 20 |
| 8.5 | 20 | - | 2.5 | - | - | <20 | 37 | <20 | 19 |
| 8.5 | - | - | 2.5 | - | - | <20 | <20 | <20 | <20 |
| 8.5 | - | - | 25 | - | - | <20 | <20 | <20 | <20 |
| 9.5 | 20 | 25 | 25 | 1 | - | 204 | 192 | >250 | 215 |
| 9.5 | - | 25 | 25 | 1 | 0.5 | 154 | 206 | 222 | 194 |
| 9.5 | - | 25 | 25 | 1 | - | 177 | 141 | 184 | 167 |
| 9.5 | 20 | 25 | 25 | - | - | 90 | 196 | 171 | 152 |
| 9.5 | - | 25 | 25 | - | - | 190 | 166 | 89 | 148 |
| 9.5 | 20 | 25 | 25 | 1 | 0.5 | 164 | 150 | 116 | 143 |
| 9.5 | 20 | - | 2.5 | 1 | 0.5 | 139 | 110 | 103 | 117 |
| 9.5 | - | 25 | 25 | - | 0.5 | 126 | 65 | 148 | 113 |
| 9.5 | 20 | 25 | 25 | - | 0.5 | 145 | 156 | <20 | 103 |
| 9.5 | - | - | 2.5 | 1 | 0.5 | 110 | 82 | 61 | 84 |
| 9.5 | 20 | - | 25 | - | 0.5 | 91 | 128 | <20 | 76 |
| 9.5 | 20 | - | 25 | 1 | 0.5 | 80 | 49 | 65 | 65 |
| 9.5 | 20 | - | 2.5 | 1 | - | 52 | 59 | 57 | 56 |
| 9.5 | - | - | 2.5 | 1 | - | 31 | 80 | 40 | 50 |
| 9.5 | - | - | 25 | 1 | 0.5 | 23 | 21 | 102 | 48 |
| 9.5 | 20 | - | 2.5 | - | - | 41 | <20 | 54 | 35 |
| 9.5 | 20 | - | 25 | - | - | <20 | 42 | 40 | 31 |
| 9.5 | 20 | - | 2.5 | - | 0.5 | <20 | <20 | 60 | 27 |
| 9.5 | - | - | 2.5 | - | 0.5 | 40 | <20 | <20 | 20 |
| 9.5 | - | - | 25 | 1 | - | 35 | <20 | <20 | 18 |
| 9.5 | - | - | 25 | - | - | <20 | 32 | <20 | 17 |
| 9.5 | - | - | 2.5 | - | - | <20 | <20 | <20 | <20 |
| 9.5 | 20 | - | 25 | 1 | - | <20 | <20 | <20 | <20 |
| 9.5 | - | - | 25 | - | 0.5 | <20 | <20 | <20 | <20 |

**Minimum detectable peak load for these analyses was 20 N*

*Ca = calcium; Mg = magnesium; Si = silicon; P = phosphorus; Zn = zinc*

*0.68 mM $HCO_3^-$ added to all tests*

*0.15 mM $SO_4^{2-}$ added to all tests*

Several trends can be observed after careful examination of the data (Table 3.2). All eight waters that were dosed with 25 mg/L magnesium required an average force greater than 100 N to break the concrete specimens apart. In addition, 9 of the 10 waters whose coupons required an average force greater than 100 N were dosed with the highest level of silicon (i.e., 25 mg/L). In combination, the results suggest that a precipitate containing magnesium might play a part in the increased strength of these specimens. Surprisingly, calcium addition was apparently not a critical factor in the development of high strength 'glues'. Of the six tests where strengths of 140 N or more were developed, calcium was added to the test water in only three and always with high magnesium and silicon. We should note, however, that due to leaching there is ample calcium present in all the test waters regardless of calcium addition.

Paired t-tests were used to determine if specific variables might be statistically significant factors in the increased strength of various concrete specimens. These were performed by choosing a subset of tests where only one variable was different. For example, there were 26 experiments in which the only difference in the water was the presence and absence of 1 mg/L orthophosphate (Figure 3.1). In 11 of 13 cases (13 waters with and without P = 26), the addition of orthophosphate increased the strength to failure, whereas in 2 of 13 cases the addition of orthophosphate decreased the strength to failure. For the statistical test, a difference data set is generated between the two sets of data and is analyzed using an assumption of a normal distribution (verified by a Shapiro-Wilk normality test ($p = 0.60$)). The result of the test indicates that the presence of phosphorus enhanced the strength gain at greater than 99% confidence ($p = 0.0047$). This approach also indicated that other constituents in the water increased strength gain including magnesium ($p = 0.00018$) and calcium ($p = 0.052$). The addition of 0.5 mg/L zinc ($p = 0.15$) or 25 mg/L silicon ($p = 0.28$) could not be proven to increase the strength with 95% confidence. One drawback of this type of analysis is that synergistic effects (such as the combination of magnesium and silicon) are not detected, but the results are nonetheless useful.

## CHEMICAL ANALYSES

Unfiltered samples taken from each container at various times (Day 1, 7, 28, 59, and 302) during the screening study were analyzed for Al, Ca, Fe, Mg, P, Si, and Zn using ICP-ES. Although aluminum and iron were not added to any of the waters tested, they were monitored to assess the magnitude of leaching which might occur. A small amount of iron (< 30 ppb in all but 2 cases) did leach from the specimens in some waters during the first week, but iron was below detection in all waters after one month. On average nearly 100 ppb aluminum (range: 52 – 272 ppb) leached from the specimens during the first day. However, by the end of the first week little or no aluminum was detected in any of the waters.

The net amount of each element leaching from the concrete specimens or precipitating from solution can be determined by examination of the concentration data obtained by ICP. Integration of the net amount of element that appeared/disappeared for each water change will result in the total gain/loss of that element for the study. If the net amount is a gain we can be certain the extra material was leached from the concrete specimens submerged in that container. On the other hand, if the net amount is a loss, we can only say that that particular element precipitated from solution.

| Test | Comparison Test | Peak Load before failure (N) without P | 1 mg/L P addition |
|---|---|---|---|
| pH 7, 25 mg/L Si and 20 mg/L Ca | plus 1 mg/L P | 43 | 28 |
| pH 9.5, 2.5 mg/L Si | plus 1 mg/L P | 10 | 50 |
| pH 9.5, 20 mg/L Ca | plus 1 mg/L P | 35 | 56 |
| pH 9.5, 20 mg/L Ca and 0.5 mg/L Zn | plus 1 mg/L P | 27 | 117 |
| pH 9.5, 25 mg/L Si | plus 1 mg/L P | 17 | 18 |
| pH 9.5, 25 mg/L Si and 20 mg/L Ca | plus 1 mg/L P | 31 | 10 |
| pH 9.5, 25 mg/L Si, 20 mg/L Ca, and 25 mg/L Mg | plus 1 mg/L P | 152 | 215 |
| pH 9.5, 25 mg/L Si, 20 mg/L Ca, and 0.5 mg/L Zn | plus 1 mg/L P | 76 | 65 |
| pH 9.5, 25 mg/L Si, 20 mg/L Ca, 25 mg/L Mg, and 0.5 mg/L Zn | plus 1 mg/L P | 103 | 143 |
| pH 9.5, 25 mg/L Si and 25 mg/L Mg | plus 1 mg/L P | 148 | 167 |
| pH 9.5, 25 mg/L Si and 0.5 mg/L Zn | plus 1 mg/L P | 10 | 48 |
| pH 9.5, 25 mg/L Si, 25 mg/L Mg, and 0.5 mg/L Zn | plus 1 mg/L P | 113 | 194 |
| pH 9.5, 0.5 mg/L Zn | plus 1 mg/L P | 20 | 84 |

| ***Data for Paired t-Test*** | *without P* | *with 1 mg/L P* |
|---|---|---|
| Mean of Peak Load before Failure (N) | 60.5 | 92.0 |
| Variance | 2715 | 4716 |
| Number of Observations | 13 | 13 |
| Degrees of Freedom | 12 | |
| t Stat (alpha = 0.05) | 3.09 | |
| P(T<=t) one-tail | 0.0047 | |
| t Critical one-tail | 1.78 | |

*Result: Since t stat > t-critical at $\alpha = 0.05$, one can be 95% confident that the mean of the '1 mg/L P addition' strengths is greater than the mean of the 'without P addition' strengths.*

**Figure 3.1 Comparison of data from tests identical in all respects except phosphorus addition**

It is uncertain whether the solid settled to the bottom of the container or formed scale on the concrete, but it is deemed most likely that it formed scale on the concrete since this is a local source of $OH^-$ which would tend to supersaturate the water.

Large amounts of calcium (range: 61 – 933 mg) were leached into solution in all but four cases (Table 3.3). These four cases were all waters at pH 9.5 and all had 20 mg/L calcium added at each water change. The concrete specimens from these four tests had fairly low strength gain (Table 3.2). In fact, for tests in which no calcium was added strength gain appears to improve as more calcium is leached into solution ($R^2 = 0.60$; Figure 3.2).

Small amounts of magnesium leached from the concrete specimens in waters to which no magnesium was added (Table 3.3). Relatively large amounts of magnesium (range: 106 – 186 mg) precipitated from waters to which 25 mg/L magnesium was originally added (indicating supersaturation with respect to the solid being formed). Silicon was leached into solution in every test, regardless of the amount dosed at each water change. As opposed to the case of calcium, strength gain appears to improve as magnesium and silicon precipitate from solution (Figures 3.3 and 3.4).

Very little zinc deposited or leached from the concrete specimens during the duration of the study and phosphorus was only measured for the first month of the study. During the first month of the study there was little change in phosphorus concentration in the water, and although problems later developed with ICP-ES measurement of phosphorus, there is no reason to believe the behavior of phosphorus would change.

## CONCLUSIONS FROM SCREENING STUDY

- Autogenous healing can be an effective method of sealing small cracks that may develop in concrete. A large 0.25 mm crack was repaired with an average strength of 215 N after being submerged in water at pH 9.5 and containing 25 mg/L Si, 20 mg/L Ca, 1 mg/L P, and 25 mg/L Mg for 12 months. Specimens submerged in other waters exhibited lower strengths.

- Water with a pH of 9.5 is more conducive to autogenous healing than water at lower pH levels.

- Autogenous healing accompanied by the regain of shear strength occurs in specimens in a number of water conditions; however, the presence of magnesium and its deposition are critical factors in whether autogenous healing, with appreciable regain of strength, will occur.

- Contrary to the current understanding of autogenous healing, calcium does not appear to be a factor in whether high strength 'glue' will form on crack surfaces.

## Table 3.3

## Estimated weight gain/loss for screening study specimens

| pH | Initial Concentration (mg/L) | | | | | Total Weight Added During Test (mg) | | | | | Estimated Elemental Weight Gain/Loss (mg)* | | | | |
|---|---|---|---|---|---|---|---|---|---|---|---|---|---|---|---|
| | Ca | Mg | Si | P | Zn | Ca | Mg | Si | P | Zn | Ca | Mg | Si | Zn | Al |
| 7 | - | - | 2.5 | - | - | - | - | 45 | - | - | -405 | -3 | -56 | 0 | 0 |
| 7 | 20 | - | 2.5 | - | - | 360 | - | 45 | - | - | -284 | -3 | -52 | 0 | 0 |
| 7 | - | - | 25 | - | - | - | - | 450 | - | - | -480 | -3 | -70 | 0 | 0 |
| 7 | 20 | - | 25 | - | - | 360 | - | 450 | - | - | -301 | -4 | -73 | 0 | 0 |
| 7 | 20 | - | 25 | 1 | - | 360 | - | 450 | 18 | - | -346 | -4 | -60 | 0 | 0 |
| 8.5 | - | - | 2.5 | - | - | - | - | 45 | - | - | -251 | -2 | -70 | 0 | 0 |
| 8.5 | 20 | - | 2.5 | - | - | 360 | - | 45 | - | - | -122 | -1 | -45 | 0 | 0 |
| 8.5 | - | - | 25 | - | - | - | - | 450 | - | - | -273 | -2 | -60 | 0 | 0 |
| 8.5 | 20 | - | 25 | - | - | 360 | - | 450 | - | - | -109 | -2 | -171 | 0 | 0 |
| 8.5 | 20 | - | 25 | - | 0.5 | 360 | - | 450 | - | 9 | -180 | -2 | -72 | 5 | 0 |
| 9.5 | - | - | 2.5 | - | - | - | - | 45 | - | - | -163 | -1 | -55 | 0 | 0 |
| 9.5 | 20 | - | 2.5 | - | - | 360 | - | 45 | - | - | *81* | 0 | -28 | 0 | 0 |
| 9.5 | 20 | - | 2.5 | 1 | - | 360 | - | 45 | 18 | - | -100 | -2 | -44 | 0 | 0 |
| 9.5 | 20 | - | 2.5 | - | 0.5 | 360 | - | 45 | - | 9 | *17* | 0 | -21 | 7 | 0 |
| 9.5 | 20 | - | 2.5 | 1 | 0.5 | 360 | - | 45 | 18 | 9 | -297 | -2 | -35 | 8 | 0 |
| 9.5 | - | - | 2.5 | 1 | - | - | - | 45 | 18 | - | -311 | -3 | -58 | 0 | 0 |
| 9.5 | - | - | 25 | - | - | - | - | 450 | - | - | -156 | -1 | -71 | 0 | 0 |
| 9.5 | 20 | - | 25 | - | - | 360 | - | 450 | - | - | *180* | 0 | -48 | 0 | 0 |
| 9.5 | 20 | 25 | 25 | - | - | 360 | 450 | 450 | - | - | -112 | 137 | -30 | 0 | 0 |
| 9.5 | 20 | - | 25 | 1 | - | 360 | - | 450 | 18 | - | -61 | -3 | -60 | 0 | 0 |
| 9.5 | 20 | 25 | 25 | 1 | - | 360 | 450 | 450 | 18 | - | -509 | 131 | -25 | 0 | 0 |
| 9.5 | 20 | - | 25 | - | 0.5 | 360 | - | 450 | - | 9 | *83* | -1 | -49 | 4 | 0 |
| 9.5 | 20 | 25 | 25 | - | 0.5 | 360 | 450 | 450 | - | 9 | -115 | 106 | -26 | 5 | 0 |
| 9.5 | 20 | - | 25 | 1 | 0.5 | 360 | - | 450 | 18 | 9 | -309 | -3 | -114 | -1 | 0 |
| 9.5 | 20 | 25 | 25 | 1 | 0.5 | 360 | 450 | 450 | 18 | 9 | -649 | 142 | -12 | 6 | 0 |
| 9.5 | - | 25 | 25 | - | - | - | 450 | 450 | - | - | -346 | 146 | -44 | 0 | 0 |
| 9.5 | - | - | 25 | 1 | - | - | - | 450 | 18 | - | -288 | -2 | -73 | 0 | 0 |
| 9.5 | - | 25 | 25 | 1 | - | - | 450 | 450 | 18 | - | -728 | 186 | -17 | 0 | 0 |
| 9.5 | - | - | 25 | - | 0.5 | - | - | 450 | - | 9 | -222 | -1 | -58 | 5 | 0 |
| 9.5 | - | 25 | 25 | - | 0.5 | - | 450 | 450 | - | 9 | -290 | 104 | -36 | 5 | 0 |
| 9.5 | - | - | 25 | 1 | 0.5 | - | - | 450 | 18 | 9 | -465 | -3 | -54 | 5 | 0 |
| 9.5 | - | 25 | 25 | 1 | 0.5 | - | 450 | 450 | 18 | 9 | -933 | 142 | -11 | 5 | 0 |
| 9.5 | - | - | 2.5 | - | 0.5 | - | - | 45 | - | 9 | -233 | -1 | -33 | 6 | 0 |
| 9.5 | - | - | 2.5 | 1 | 0.5 | - | - | 45 | 18 | 9 | -445 | -3 | -48 | 7 | 0 |

** Estimated Elemental Weight Gain/Loss' is calculated from concentrations added to each container at each water change and concentrations at the end of each water change as determined by ICP-ES analyses; e.g., Mg wt gain = container volume * summation of (Mg dosed to container - Mg remaining at water change) for each water change*

*(-) indicates weight loss [i.e. element leached from specimens]*

*Ca = calcium; Mg = magnesium; Si = silicon; P = phosphorus; Zn = zinc; Al = aluminum*

*0.68 mM $HCO_3^-$ added to all tests*

*0.15 mM $SO_4^{2-}$ added to all tests*

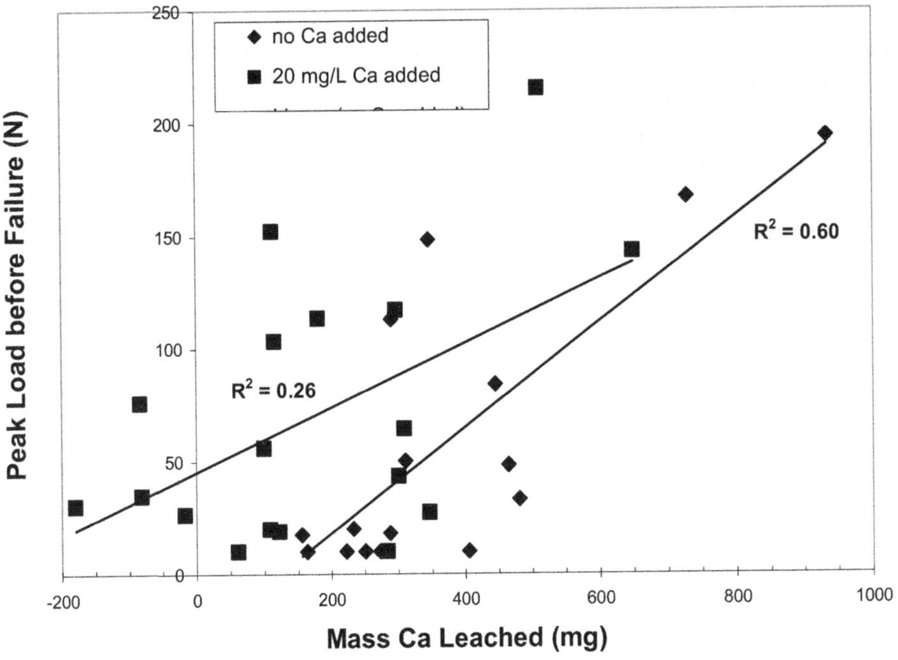


**Figure 3.2 Peak load before failure as a function of mass of calcium leached from specimens. Surprisingly, strength improves as more calcium is lost from specimens.**

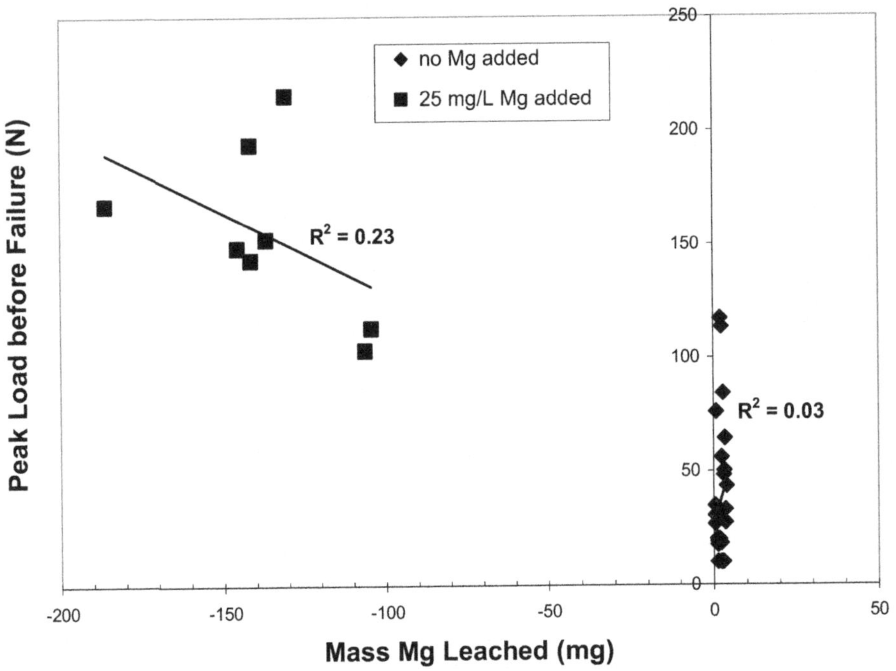


**Figure 3.3 Peak load before failure as a function of mass of magnesium leached from specimens. Strength possibly increases as more magnesium precipitates from solution.**

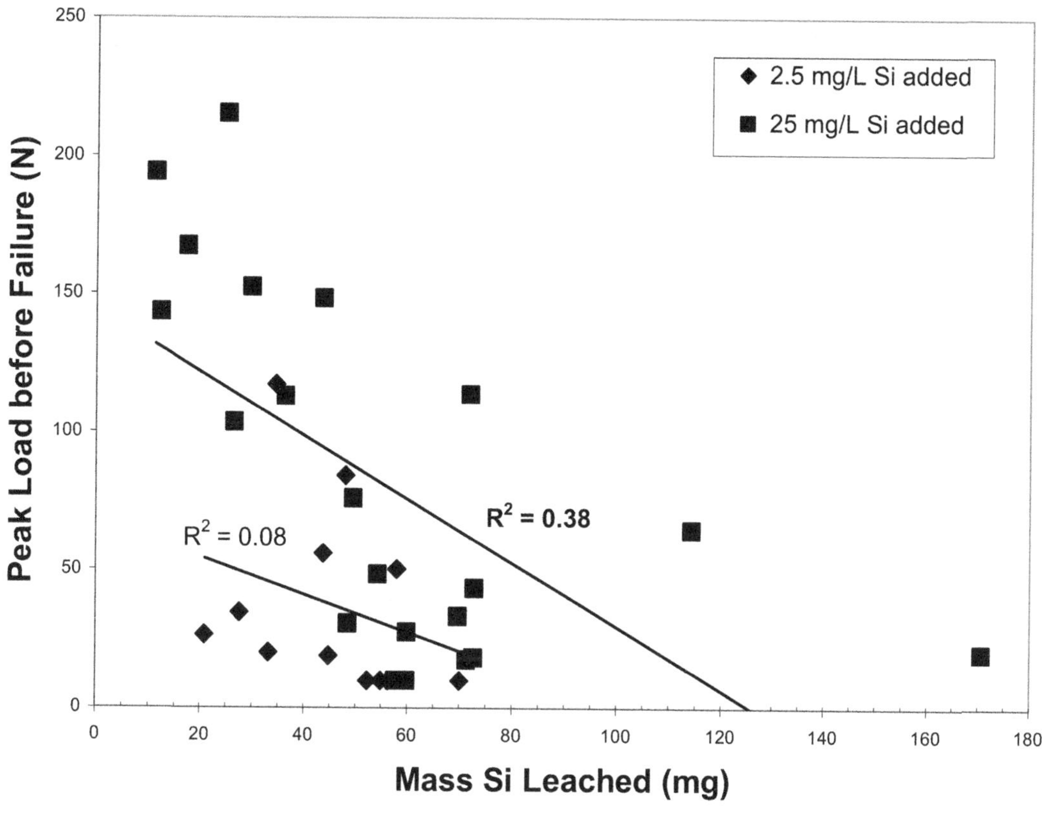


**Figure 3.4 Peak load before failure as a function of mass of silicon leached from specimens. Strength gain appears to be inversely proportional to amount of silicon lost from specimens.**

# CHAPTER 4: CHARACTERIZATION OF THE EFFECTS OF WATER CHEMISTRY ON CONCRETE CORROSION AND AUTOGENOUS REPAIR

The results from the initial screening study (Chapter 3) provided evidence that water chemistry is a critical parameter in determining the extent to which autogenous healing will occur. Many questions were not answered, however, so an in-depth study was conducted by making adjustments to the most promising water chemistries from the screening study.

## SELECTION OF PARAMETERS FOR IN-DEPTH STUDY

The primary questions that required answers at the conclusion of the screening study were:

1- *What role does silicon play in autogenous healing?* Data from the screening study indicated that the presence of magnesium enhances the strength of the glue formed during autogenous healing. However, all tests that were conducted with magnesium also had 25 mg/L silicon present in the water. Therefore, the possibility existed that silicon may also contribute towards autogenous healing as well. The in-depth study included tests with magnesium to address whether magnesium or the combination of magnesium and silicon is necessary for appreciable autogenous healing to occur.

2- *Can calcium carbonate precipitation cause autogenous healing if higher levels of calcium and bicarbonate are introduced?* Many researchers have asserted that calcium carbonate precipitation is the primary mechanism behind autogenous healing (Table 1.1). Although high levels of calcium leached from specimens in the screening study, it is quite possible the 0.68 mM total carbonate was not sufficient to induce high strength healing. Higher levels of carbonate were added in some tests in the in-depth study to address this question.

3- *Can phosphorus and/or zinc be responsible for autogenous healing (without the presence of magnesium or silicon)?* Literature reviews (AWWA 2001) have suggested that phosphorus or zinc inhibitors can be protective to concrete. Schock and Buelow (1981) suggest that zinc may precipitate from solution under the right conditions to form a zinc hydroxycarbonate scale on the concrete surface. However, there is nothing definitive in the literature proving cause, effect or mechanism of such protection, if indeed it is significant.

4- *What effect does natural organic matter (NOM) have on autogenous healing?* Preliminary research has concluded that NOM is often a component of protective films formed on concrete (AwwaRF and DVGW-TZW 1996). The in-depth study included several tests to which NOM was added to evaluate its effect.

Concentrations of calcium, magnesium, silicon, zinc, phosphorus, NOM, and bicarbonate were selected to address each of these questions (Table 4.1).

**Table 4.1**

**Composition of test waters for in-depth study**

| pH | Initial Concentration (mg/L) | | | | | | $HCO_3^-$ |
|---|---|---|---|---|---|---|---|
| | Ca | Mg | Si | P | Zn | NOM | (mM) |
| 7 | - | - | - | - | - | - | 0.1 |
| 7 | 20 | 25 | 25 | 1 | - | - | 0.1 |
| 8 | - | - | - | - | - | - | 0.1 |
| 8 | 20 | 25 | 25 | 1 | - | - | 0.1 |
| 8 | 45 | - | - | - | - | - | 4 |
| 8 | 70 | - | - | - | - | - | 10 |
| 8 | - | 10 | 20 | - | - | - | 0.1 |
| 8 | - | 10 | 35 | - | - | - | 0.1 |
| 8 | - | 40 | - | - | - | - | 0.1 |
| 8 | - | 40 | 20 | - | - | - | 0.1 |
| 8 | - | 40 | 35 | - | - | - | 0.1 |
| 9.5 | - | - | - | - | - | - | 0.1 |
| 9.5 | 20 | 25 | 25 | 1 | - | - | 0.1 |
| 9.5 | 45 | - | - | - | - | - | 4 |
| 9.5 | 70 | - | - | - | - | - | 10 |
| 9.5 | 70 | - | - | - | - | 5 | 10 |
| 9.5 | - | 10 | 20 | - | - | - | 0.1 |
| 9.5 | - | 10 | 35 | - | - | - | 0.1 |
| 9.5 | - | 40 | - | - | - | - | 0.1 |
| 9.5 | - | 40 | 20 | - | - | - | 0.1 |
| 9.5 | - | 40 | 35 | - | - | - | 0.1 |
| 9.5 | - | 40 | 35 | - | - | 5 | 0.1 |
| 9.5 | - | - | - | 3 | - | - | 0.1 |
| 9.5 | - | - | - | 3 | 1 | - | 0.1 |
| 9.5 | - | - | - | 3 | 1 | 5 | 0.1 |
| 9.5 | - | - | - | - | 1 | - | 0.1 |

*Ca = calcium; Mg = magnesium; Si = silicon; P = phosphorus; Zn = zinc; NOM = natural organic matter;* $HCO_3^-$ *= bicarbonate*

*0.15 mM* $SO_4^{2-}$ *added to all tests*

## VISUAL OBSERVATIONS

A more extensive set of visual observations was made during the in-depth study (Table 4.2). At Day 42 only two of the 78 specimens had a visible precipitate protruding from the crack surface; however, no light could be seen through the crack of 29 of the 78 specimens. The lack of visible light indicates that some material must be sealing the crack even if no precipitate is readily apparent. By the end of the test on Day 232 six of the 78 specimens had a visible precipitate.

Visual observation demonstrated that in all of the water chemistries tested at least *some* crack closure occurred (selected digital images in Appendix B). While this observation seems to corroborate the findings of Edvarsen (1999) who concluded that water chemistry had little effect on the ability of concrete to heal itself, some surprising observations were noted. Of the four test conditions where the average crack was still at least 90% visible after Day 232, three of these were tests in which calcium was added to the water. This observation appears to be in direct contrast to the assertion that calcium carbonate precipitation may be the primary mechanism behind autogenous healing, since higher calcium increases the likelihood and extent of calcium carbonate precipitation. It is also of interest to note that replicates in the same container do not necessarily behave similarly. For example, the test at pH 9.5 with 1 mg/L Zn had one specimen that was completely sealed with a visible white precipitate and another specimen with 100% of its crack still visible after Day 232. These disparate behaviors from specimens in the same water reservoir indicate differences resulting from one coupon to another and/or placement within the container.

## CHEMICAL ANALYSES

Water samples taken from each container prior to each water change were analyzed for Al, Ca, Mg, P, Si, and Zn using ICP-MS. As in the screening study, the net amount of each element leaching from the concrete specimens or precipitating from solution can be determined by examination of the concentration data obtained by ICP (Table 4.3). Each element of interest will be discussed in the paragraphs that follow.

### Calcium

As expected, calcium leached into solution for all water conditions except for the 5 tests to which 45 or 70 mg/L calcium was added (Table 4.3). The highest level of calcium (515 mg) leached from the pH 8 water with 40 mg/L Mg and no other added constituents. While this value may appear to be large for a 7 month test, it is equivalent to only 1.3% of the original combined weight of the three specimens in this container. The soluble calcium concentrations at the end of each water change cycle illustrate the degree to which leaching occurs (Figure 4.1). At pH 9.5, calcite precipitation controls calcium solubility to levels below 1 mg/L in most instances. NOM addition increased calcium levels to about 2.4 mg/L (a 240% increase over comparable water without NOM). Calcite still controls calcium solubility at pH 8, however at much higher levels than at pH 9.5. Data show that the amount of calcium in solution at pH 8 is highly dependent on carbonate concentration. As more carbonate was dosed, the steady state soluble calcium concentration decreased.

**Table 4.2**
**Visual observations of specimens from in-depth study (3 per solution)**

| pH | Initial Concentration (mg/L) | | | | | | | Number of Specimens with Visible Crack | | | | | | Day 232: % crack still visible | | | | |
|---|---|---|---|---|---|---|---|---|---|---|---|---|---|---|---|---|---|---|
| | Ca | Mg | Si | P | Zn | NOM | $HCO_3^-$ (mM) | Day 42 | Day 70 | Day 102 | Day 140 | Day 180 | Day 224 | a* | b | c | | Avg |
| 7 | - | - | - | - | - | - | 0.1 | 3 | 3 | 3 | 3 | 3 | 2 | 50 | 40 | 0 | - | 30 |
| 7 | 20 | 25 | 25 | 1 | - | - | 0.1 | 2 | 2 | 2 | 2 | 2 | 3 | 10 | 10 | 80 | - | 33 |
| 8 | - | - | - | - | - | - | 0.1 | 0 | 3 | 3 | 1 | 0 | 2 | 40 | 0 | 90 | - | 43 |
| 8 | 20 | 25 | 25 | 1 | - | - | 0.1 | 2 | 3 | 3 | 3 | 3 | 3 | 30 | 30 | 100 | - | 53 |
| 8 | 45 | - | - | - | - | - | 4 | 3 | 2 | 3 | 3 | 3 | 3 | 10 | 70 | 40 | - | 40 |
| 8 | 70 | - | - | - | - | - | 10 | 3 | 3 | 3 | 3 | 3 | 3 | 100 | 80 | 90 | - | 90 |
| 8 | - | 10 | 20 | - | - | - | 0.1 | 3 | 3 | 3 | 3 | 0 | 3 | 100 | 100 | 80 | - | 93 |
| 8 | - | 10 | 35 | - | - | - | 0.1 | 0 | 1 | 3 | 3 | 2 | 2 | 60 | 10 | 0 | - | 23 |
| 8 | - | 40 | - | - | - | - | 0.1 | 2 | 3 | 3 | 2 | 2 | 3 | 30 | 10 | 10 | - | 17 |
| 8 | - | 40 | 20 | - | - | - | 0.1 | 0 | 3 | 3 | 3 | 3 | 3 | 30 | 30 | 10 | - | 23 |
| 8 | - | 40 | 35 | - | - | - | 0.1 | 2 | 3 | 3 | 3 | 1 | 1 | 0 | 0 | 100 | - | 33 |
| 9.5 | - | - | - | - | - | - | 0.1 | 3 | 3 | 3 | 3 | 3 | 3 | 100 | 50 | 100 | - | 83 |
| 9.5 | 20 | 25 | 25 | 1 | - | - | 0.1 | 1 | 1 | 2 | 1 | 2 | 1 | 0 | 0 | 30 | - | 10 |
| 9.5 | 45 | - | - | - | - | - | 4 | 3 | 3 | 3 | 3 | 3 | 3 | 90 | 90 | 90 | - | 90 |
| 9.5 | 70 | - | - | - | - | - | 10 | 3 | 3 | 3 | 3 | 3 | 3 | 70 | 100 | 100 | - | 90 |
| 9.5 | 70 | - | - | - | - | 5 | 10 | 3 | 3 | 3 | 3 | 3 | 3 | 80 | 70 | 100 | - | 83 |
| 9.5 | - | 10 | 20 | - | - | - | 0.1 | 0 | 0 | 3 | 2 | 0 | 1 | 0 | 0 | 10 | - | 3 |
| 9.5 | - | 10 | 35 | - | - | - | 0.1 | 3 | 3 | 3 | 3 | 3 | 3 | 100 | 100 | 10 | - | 70 |
| 9.5 | - | 40 | - | - | - | - | 0.1 | 0 | 1 | 3 | 2 | 2 | 1 | 0 | 10 | 0 | - | 3 |
| 9.5 | - | 40 | 20 | - | - | - | 0.1 | 2 | 2 | 2 | 2 | 0 | 2 | 90 | ***0*** | 20 | - | 37 |
| 9.5 | - | 40 | 35 | - | - | - | 0.1 | 0 | 0 | 0 | 0 | 0 | 0 | ***0*** | ***0*** | 0 | - | 0 |
| 9.5 | - | 40 | 35 | - | - | 5 | 0.1 | 0 | 0 | 0 | 0 | 0 | 0 | 0 | 0 | 0 | - | 0 |
| 9.5 | - | - | - | 3 | - | - | 0.1 | 3 | 2 | 2 | 2 | 2 | 2 | 0 | 100 | 100 | - | 67 |
| 9.5 | - | - | - | 3 | 1 | - | 0.1 | 3 | 3 | 3 | 0 | 1 | 2 | 70 | ***20*** | ***0*** | - | 30 |
| 9.5 | - | - | - | 3 | 1 | 5 | 0.1 | 2 | 0 | 0 | 1 | 0 | 2 | 20 | 0 | 30 | - | 17 |
| 9.5 | - | - | - | - | 1 | - | 0.1 | 3 | 3 | 3 | 3 | 2 | 2 | 20 | ***0*** | 100 | - | 40 |

*Ca = calcium; Mg = magnesium; Si = silicon; P = phosphorus; Zn = zinc; NOM = natural organic matter; HCO3- = bicarbonate*
*0.15 mM* $SO_4^{2-}$ *added to all tests*
* *'a', 'b', and 'c' indicate code used to identify specimens taken from containers*
*Boxed data indicate specimen had visible precipitate protruding from crack surface*

## Table 4.3

## Elemental weight gain/loss for in-depth study specimens

| pH | Initial Concentration (mg/L) | | | | | | Initial $HCO_3^-$ (mM) | Elemental Weight Gain/Loss (mg)* | | | | | |
|---|---|---|---|---|---|---|---|---|---|---|---|---|---|
| | Ca | Mg | Si | P | Zn | NOM | | Ca | Mg | Si | P | Zn | Al |
| 7 | - | - | - | - | - | - | 0.1 | 454.9 | 4.6 | 71.4 | 0.1 | 0.0 | 1.9 |
| 7 | 20 | 25 | 25 | 1 | - | - | 0.1 | 450.1 | -86.5 | -40.2 | -3.3 | 0.0 | 0.1 |
| 8 | - | - | - | - | - | - | 0.1 | 396.4 | 3.5 | 53.9 | 0.1 | 0.0 | 4.1 |
| 8 | 20 | 25 | 25 | 1 | - | - | 0.1 | 294.8 | -91.4 | -1.0 | -2.2 | 0.0 | 0.2 |
| 8 | 45 | - | - | - | - | - | 4 | -303.9 | 1.3 | 10.0 | 0.1 | 0.0 | 1.8 |
| 8 | 70 | - | - | - | - | - | 10 | -965.7 | 1.0 | 9.5 | 0.1 | 0.0 | 1.3 |
| 8 | - | 10 | 20 | - | - | - | 0.1 | 453.0 | -66.7 | -27.7 | 0.1 | 0.0 | 0.2 |
| 8 | - | 10 | 35 | - | - | - | 0.1 | 470.0 | -66.3 | -63.3 | 0.2 | 0.0 | 0.1 |
| 8 | - | 40 | - | - | - | - | 0.1 | 515.1 | -97.2 | 31.8 | 0.1 | 0.0 | 1.8 |
| 8 | - | 40 | 20 | - | - | - | 0.1 | 479.8 | -124.8 | -35.9 | 0.2 | 0.0 | 0.1 |
| 8 | - | 40 | 35 | - | - | - | 0.1 | 485.7 | -143.9 | -77.8 | 0.2 | 0.0 | 0.1 |
| 9.5 | - | - | - | - | - | - | 0.1 | 129.1 | 2.2 | 60.5 | 0.2 | 0.0 | 6.0 |
| 9.5 | 20 | 25 | 25 | 1 | - | - | 0.1 | 231.4 | -79.4 | -48.0 | -3.9 | 0.0 | 0.1 |
| 9.5 | 45 | - | - | - | - | - | 4 | -689.6 | 0.9 | 11.6 | 0.1 | 0.0 | 1.7 |
| 9.5 | 70 | - | - | - | - | - | 10 | -1104.4 | 0.8 | 15.4 | 0.1 | 0.0 | 3.0 |
| 9.5 | 70 | - | - | - | - | 5 | 10 | -1075.4 | 2.1 | 28.0 | 0.7 | 0.0 | 3.4 |
| 9.5 | - | 10 | 20 | - | - | - | 0.1 | 253.5 | -72.1 | -42.3 | 0.1 | 0.0 | 0.3 |
| 9.5 | - | 10 | 35 | - | - | - | 0.1 | 261.2 | -87.8 | -93.0 | 0.1 | 0.0 | 0.3 |
| 9.5 | - | 40 | - | - | - | - | 0.1 | 278.1 | -88.5 | 21.9 | 0.1 | 0.0 | 0.4 |
| 9.5 | - | 40 | 20 | - | - | - | 0.1 | 288.2 | -157.7 | -71.7 | 0.2 | 0.0 | 0.1 |
| 9.5 | - | 40 | 35 | - | - | - | 0.1 | 308.2 | -243.4 | -207.3 | 0.1 | 0.0 | 0.2 |
| 9.5 | - | 40 | 35 | - | - | 5 | 0.1 | 346.2 | -163.4 | -108.5 | 1.7 | 0.2 | 0.6 |
| 9.5 | - | - | - | 3 | - | - | 0.1 | 188.9 | 2.4 | 48.2 | -27.6 | 0.0 | 3.2 |
| 9.5 | - | - | - | 3 | 1 | - | 0.1 | 208.6 | 2.4 | 49.3 | -22.7 | -9.9 | 3.2 |
| 9.5 | - | - | - | 3 | 1 | 5 | 0.1 | 272.9 | 2.8 | 51.1 | -10.1 | -2.9 | 3.1 |
| 9.5 | - | - | - | - | 1 | - | 0.1 | 226.9 | 2.4 | 53.1 | 0.1 | -14.0 | 4.1 |

*Ca = calcium; Mg = magnesium; Si = silicon; P = phosphorus;*
*Zn = zinc; NOM = natural organic matter;* $HCO_3^-$ *= bicarbonate*
*0.15 mM* $SO_4^{2-}$ *added to all tests*

**Calculated by Summation (final concentration - initial concentration) * 2 liter [for each water change]*
*Negative (-) indicates element was lost from solution; i.e. via precipitation*
*Positive (+) indicates element was leached from concrete specimens*

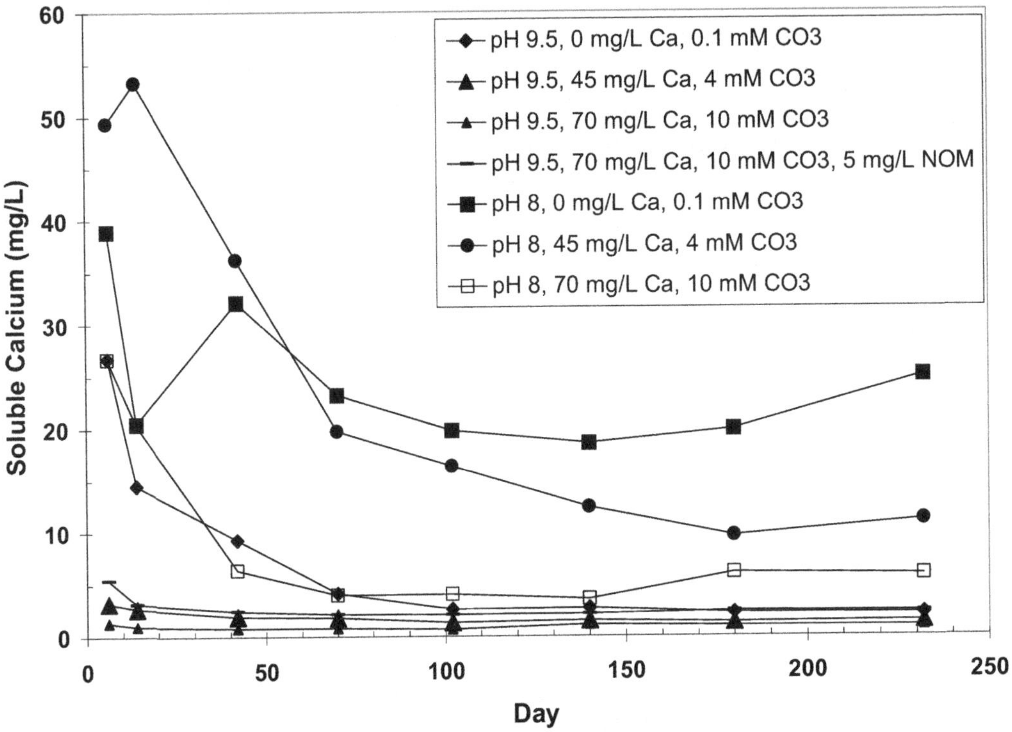


**Figure 4.1 Soluble calcium concentrations prior to each water change (data shown for tests in which calcium and bicarbonate were added).**

Although only 0.1 mM carbonate was dosed into the control there is a possibility that small amounts of carbonate may leach from the concrete specimens. In summary, if hindering dissolution of calcium from concrete is a primary distribution system goal, high pH is critical unless a high concentration of carbonate (i.e. alkalinity) is present in the water.

## Magnesium and Silicon

Magnesium and/or silicon were added to several test waters to determine their role in autogenous healing. In all waters to which it was added, magnesium was removed from solution. As in the screening study (Chapter 3) a precipitate formed at the crack of several specimens that were submerged in waters with high levels of magnesium and silicon (Figure 4.2). However, further examination of soluble calcium concentration data suggests that even though precipitates may be forming on concrete surfaces, overall calcium is still leaching from the specimens. The addition of 40 mg/L magnesium to pH 8 waters with 0, 20, or 35 mg/L silicon resulted in higher release of calcium from specimens (Figure 4.3). In pH 9.5 waters the effect was more pronounced (Figure 4.4). If magnesium were forming a protective coating on the concrete, one would expect that the extent of calcium leaching to water would decrease. Therefore, we speculate that magnesium may be participating in an ion-exchange reaction with surface calcium in these instances, since additional magnesium in the water enhanced loss of calcium. No tests were conducted with silicon in the absence of magnesium.

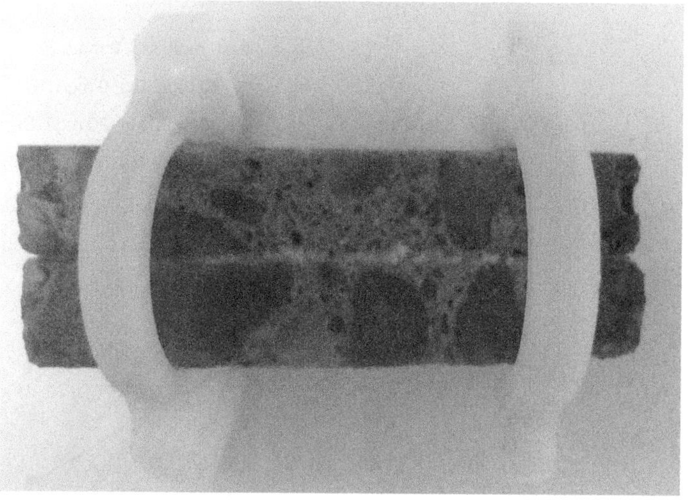

**Figure 4.2 Close-up of specimen after 7 months in pH 9.5, 40 mg/L Mg, 20 mg/L Si water. Note crystalline precipitate protruding from crack.**

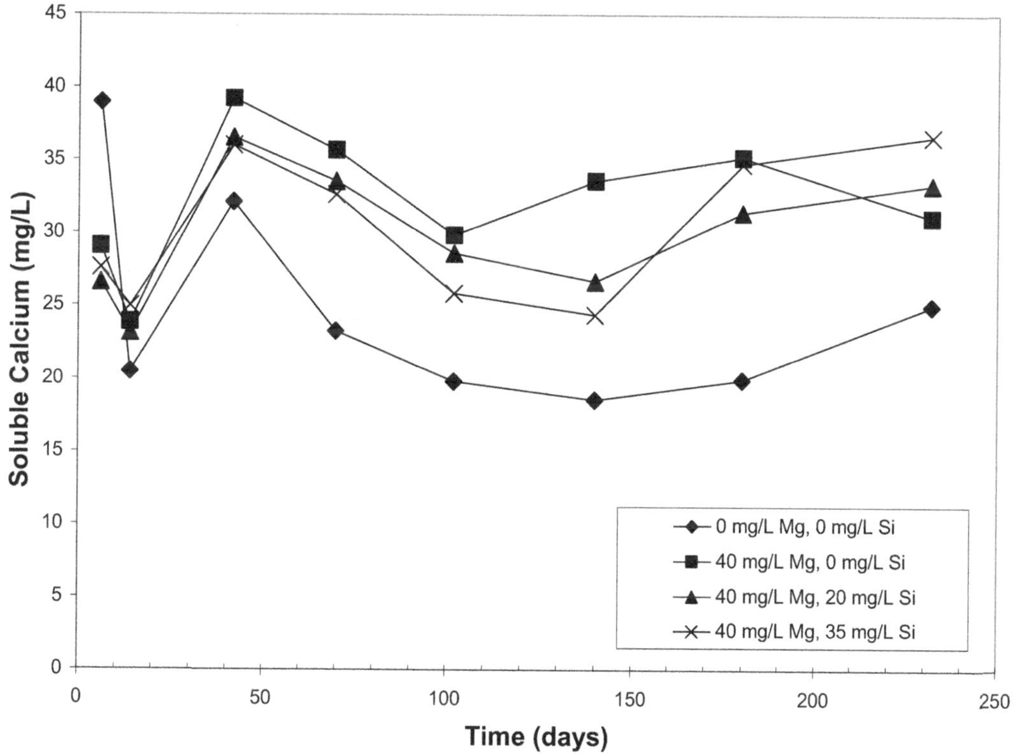


**Figure 4.3 Soluble calcium concentrations at each water change. All tests conducted at pH 8 with 0.1 mM total carbonate.**

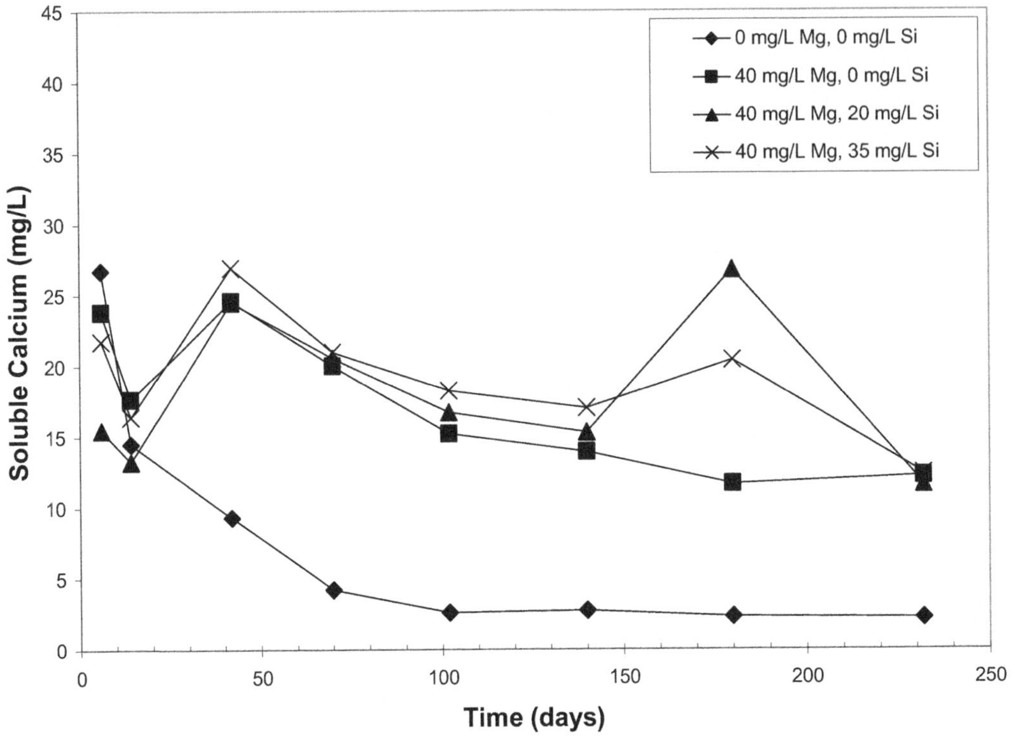


**Figure 4.4 Soluble calcium concentrations at each water change. All tests conducted at pH 9.5 with 0.1 mM total carbonate.**

### Zinc and Orthophosphate

Zinc and/or orthophosphate were added to several test waters to determine their role in autogenous healing and concrete corrosion inhibition. These were tested because of their frequent use as corrosion inhibitors (either as zinc orthophosphate or as orthophosphate alone) by water treatment utilities. Three tests were performed with orthophosphate in bulk water at pH 9.5 and 0.1 mM bicarbonate. In each test the initial concentration of phosphorus was approximately 3 mg/L (as P). One container was dosed with 1 mg/L zinc (e.g., zinc phosphate) and the third container was dosed with 1 mg/L zinc and 5 mg/L NOM (as C).

In the case of phosphorus addition alone or in conjunction with only zinc, soluble phosphorus was deposited on the concrete surface as indicated by decreased concentration in the bulk water (Figure 4.5). For the test with 5 mg/L NOM addition, soluble phosphorus remained at a fairly consistent level of approximately 2.5 mg/L.

Three tests were also performed with zinc in bulk water at pH 9.5 and 0.1 mM bicarbonate. Two of these tests were described above. The third contained only 1 mg/L zinc without phosphorus or NOM addition. As in the case of phosphorus, soluble zinc was depositing on the concrete surface, but to a much lesser extent in the presence of NOM (Figure 4.6). This indicates that NOM exerts a negative influence on deposition of zinc and phosphorus to concrete surfaces.

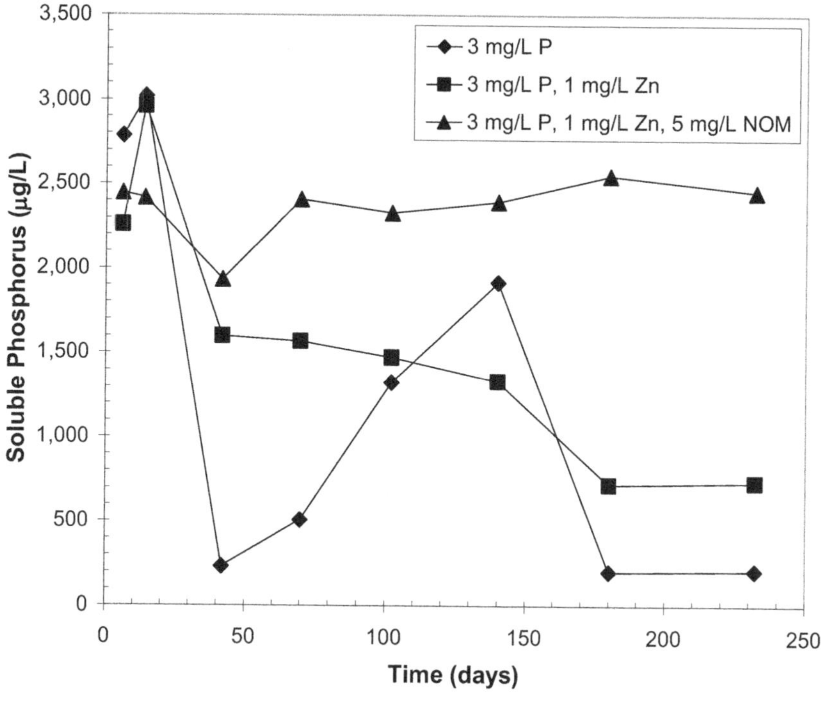


**Figure 4.5 Soluble phosphorus concentrations at each water change. All tests conducted at pH 9.5 with 0.1 mM total carbonate.**

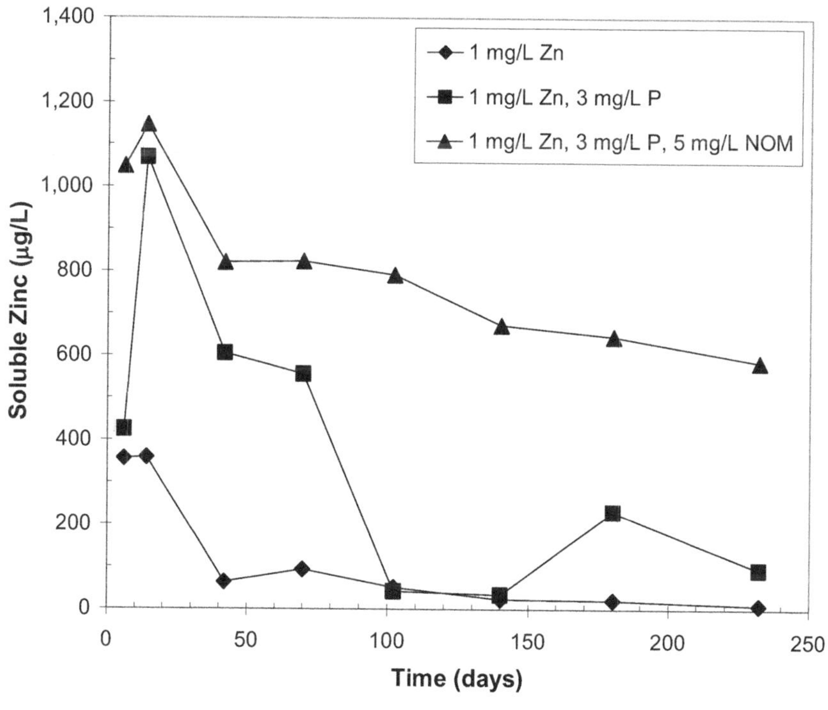


**Figure 4.6 Soluble zinc concentrations at each water change. All tests conducted at pH 9.5 with 0.1 mM total carbonate.**

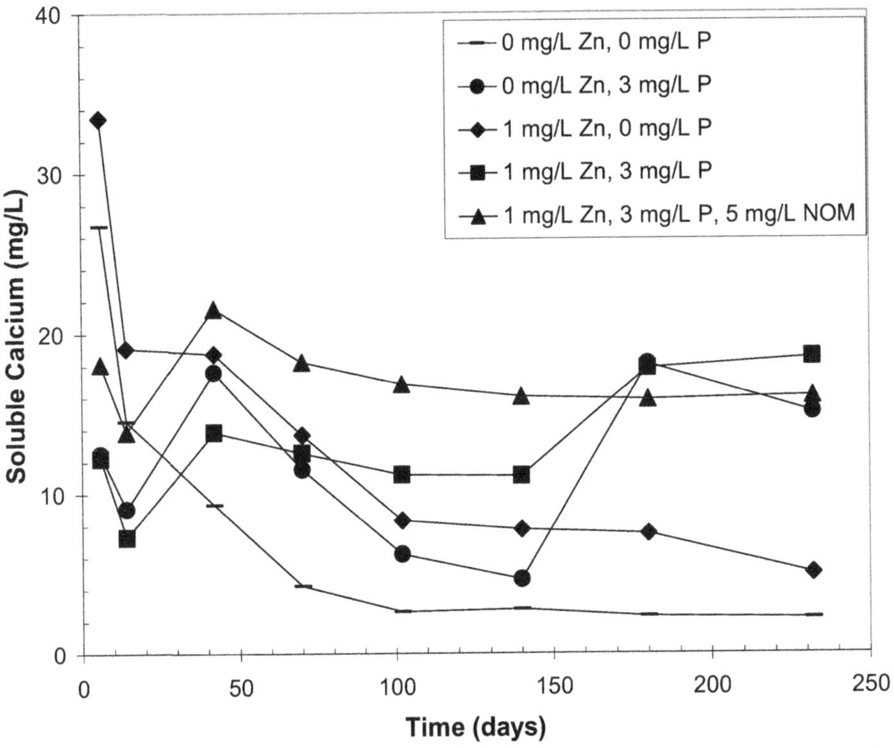


**Figure 4.7 Soluble calcium concentrations at each water change. All tests conducted at pH 9.5 with 0.1 mM total carbonate. Surprisingly, less calcium is leached from coupons submerged in a 'control' solution than from solutions containing 3 mg/L P and/or 1 mg/L Zn.**

In addition to the tests described above for phosphorus and/or zinc addition, a 'control' was also tested. This test was conducted at pH 9.5 with 0.1 mM bicarbonate; however, no phosphorus, zinc, or NOM was added. In all cases, soluble calcium concentrations throughout the test period were higher than in the control for all conditions (Figure 4.7), demonstrating that the addition of phosphorus or zinc actually caused more calcium to leach into the bulk water. Thus, it is clear that in at least some cases, phosphate and zinc can actually enhance corrosion of concrete as measured by calcium leaching that likely resulted from lime dissolution.

**Aluminum**

Aluminum was not added to any of the test waters, but was still monitored because it is a major constituent in concrete/cement (portland cement is 4.5% $Al_2O_3$ by weight) and there is evidence that it can leach from concrete in some instances (Berend and Trouwborst 1999). Aluminum is not currently regulated by the Safe Drinking Water Act as a primary contaminant; however, a secondary maximum contaminant level range of 50 – 200 ppb has been established to ensure removal of coagulated solids at the drinking water treatment plant prior to release of water to the distribution system.

Thirteen of the 26 tests had average soluble aluminum concentrations that exceeded the lower 50 ppb SMCL and 5 of these tests actually exceeded the 200 ppb SMCL (Figure 4.8).

Magnesium and/or silicon appeared to play a vital role in aluminum leaching. All 13 tests that exceeded 50 ppb aluminum had no silicon in the test water. This result corroborates the findings of Slats (2002) who discovered aqueous silicates decreased aluminum leaching. Twelve of the 13 tests with high aluminum occurred in the waters without added magnesium, consistent with expectations based on Kvech and Edwards (2002) that an alumino-magnesium silicate solid may be forming in the concrete pores, thereby reducing the amount of aluminum leaching into the bulk water.

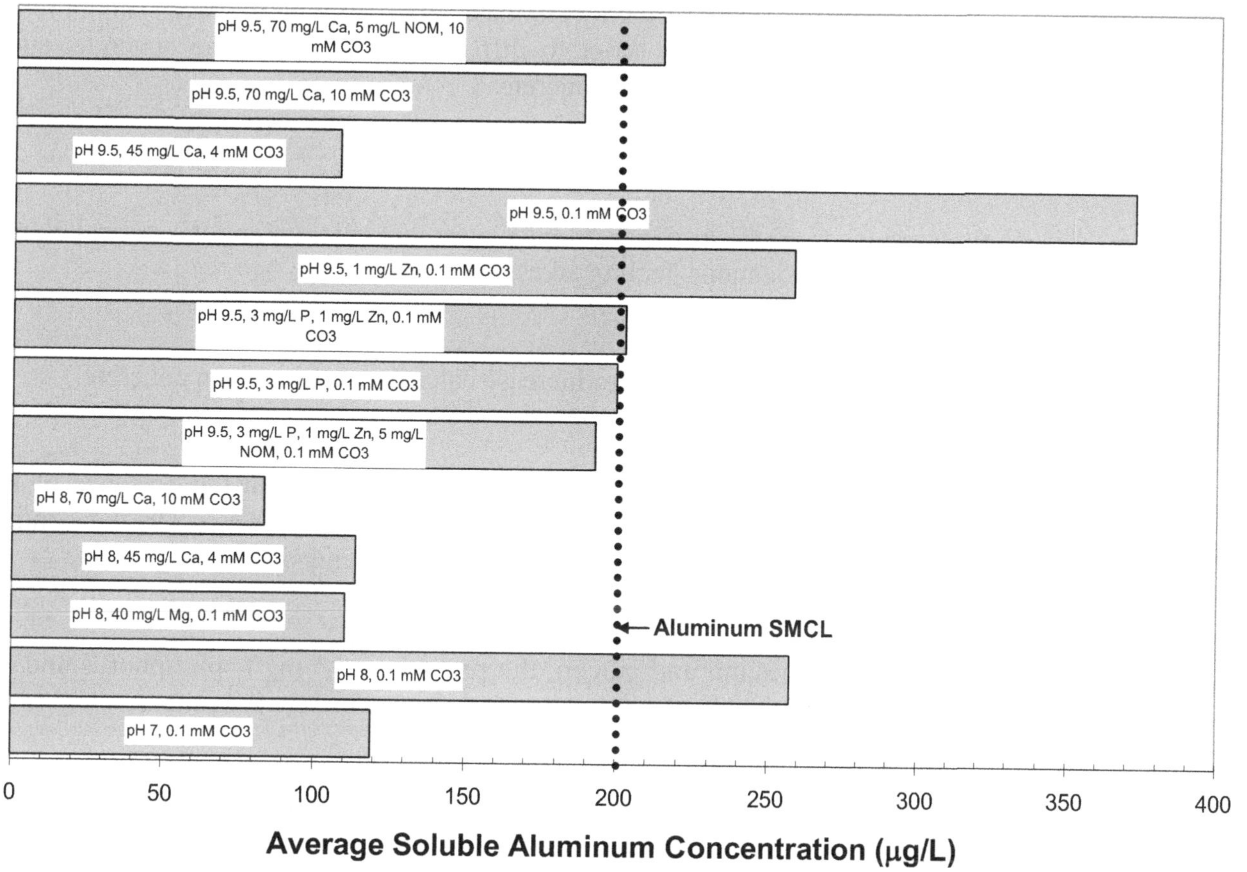


**Figure 4.8 Average soluble aluminum concentrations from in-depth study tests. (Only tests with soluble aluminum greater than 50 μg/L are shown.)**

## CARBONATION TESTING

As noted previously, carbonation occurs as a result of attack by carbon dioxide and the lower pore water pH ultimately causes corrosion of underlying steel in reinforced concrete. Carbonation testing conducted on 14 specimens demonstrated that carbonation of interior concrete surfaces was inhibited in some of the water conditions studied (selected digital images in Appendix C). Of the 8 specimens with 0 – 10% of the initial crack still visible at the end of the study, none showed signs of carbonation occurring on the internal crack surfaces. Four of the 6 specimens with >30% of the initial crack still visible tested positive for carbonation. Concrete specimens submerged in the same container did not always show similar results. Closure of the crack however, whether any strength was present or not, appeared to prevent carbonation of the inner surfaces. This implies that while calcium carbonate solids forming in cracks cannot restore strength, they do serve as an effective barrier to diffusion of $CO_2$ into the concrete, and/or calcium and lime diffusion outwards from the concrete.

## CONCLUSIONS

- Visual observations from the in-depth study suggest that calcium is not a critical element in affecting whether autogenous healing takes place.

- The presence of NOM in water tends to increase calcium leaching from concrete.

- The formation of a precipitate on concrete surfaces at pH 9.5 in the presence of high levels of magnesium and silicon does not inhibit calcium leaching. On the contrary, soluble calcium increases when high levels of magnesium and silicon are present.

- As in the case of magnesium and silicon, the presence of 3 mg/L phosphorus and/or 1 mg/L zinc causes higher levels of calcium to leach from concrete specimens.

- The presence of silicon controls leaching of aluminum from portland cement in a range that is significant relative to the current SMCL.

- Carbonation of internal concrete surfaces was not detected when cracks were sealed via autogenous healing.

# CHAPTER 5: RESULTS FROM MECHANICAL AND CHEMICAL TESTING OF AUTOGENOUSLY HEALED CRACKS

In Chapter 4 evidence was presented to demonstrate how water chemistry could affect the degree of autogenous healing of concrete specimens. Visual evidence illustrated that in all water chemistries tested some autogenous healing occurred. In this chapter we will present data from mechanical tests (strength testing) and chloride diffusion tests to assess the extent to which autogenous healing occurred.

## STRENGTH TESTING

Strength testing was conducted on all specimens from the in-depth study in a manner similar to that employed during the screening study. The protocol used on the strength testing device was changed slightly due to a problem with the instrument (as discussed in Chapter 2). This new protocol however, turned out to be more useful in that we could measure strengths lower than 20 N.

Specimens from the in-depth study did not achieve the strengths observed in the screening study (Table 5.1). Although the water chemistry was not identical in any containers, the highest strength condition (pH 9.5 with 25 mg/L Si, 25 mg/L Mg, 20 mg/L Ca, and 1 mg/L P) from the screening study was repeated (with only a slight difference in bicarbonate concentration) to check for consistency (Note: The pH 7 control test and the pH 9.5 control test were also identical except for bicarbonate concentration). Results for this high strength condition based on the screening study results were completely different in the retest – the average strength gain was only 2 N compared to 215 N in the screening study. Several factors could have affected these results. The most likely is the test duration. The in-depth study was conducted for only 7 months compared to 12 months for the screening study. Other factors which differed include the frequency of pH adjustment and the operation of the orbital shakers. During the final 5 months of the screening study, pH adjustment was performed less frequently than during the in-depth study, and a breakdown of the orbital shaker resulted in a long period of stagnation. These factors of stagnation and higher pH might have enhanced the rate of autogenous repair, or the extra time of the study might have been the critical factor.

In any event, comparisons could still be made between water conditions within each study. In a manner similar to the screening study, the four highest strength conditions all contained 40 mg/L magnesium and at least 20 mg/L silicon in the in-depth study (Table 5.1). This further supports the notion that a magnesium silicate precipitate might play a part in the increased strength of these specimens.

**Table 5.1**

**Strength testing results for in-depth study (sorted by strength within each pH block)**

| pH | Initial Concentration (mg/L) | | | | | | $HCO_3^-$ | Peak Load (N) before failure | | | |
|---|---|---|---|---|---|---|---|---|---|---|---|
| | | | | | | | | Specimen # | | | |
| | Ca | Mg | Si | P | Zn | NOM | (mM) | 1 | 2 | 3 | Avg. |
| 7 | 20 | 25 | 25 | 1 | - | - | 0.1 | 11.2 | 5.7 | 2.5 | 6.5 |
| 7 | - | - | - | - | - | - | 0.1 | 0 | 0 | 1.5 | 0.5 |
| 8 | - | 40 | 35 | - | - | - | 0.1 | 49.6 | 47.8 | 11.3 | 36.2 |
| 8 | - | 40 | 20 | - | - | - | 0.1 | 14.9 | 9.1 | 37.8 | 20.6 |
| 8 | 45 | - | - | - | - | - | 4 | 11 | 8.1 | 18.1 | 12.4 |
| 8 | - | 10 | 35 | - | - | - | 0.1 | 1.4 | 5.9 | 26.6 | 11.3 |
| 8 | - | 40 | - | - | - | - | 0.1 | 26.8 | 0 | 6.7 | 11.2 |
| 8 | - | - | - | - | - | - | 0.1 | 7.3 | 0 | 1.5 | 2.9 |
| 8 | - | 10 | 20 | - | - | - | 0.1 | 1.4 | 0 | 1.6 | 1.0 |
| 8 | 70 | - | - | - | - | - | 10 | 0 | 0 | 0.9 | 0.3 |
| 8 | 20 | 25 | 25 | 1 | - | - | 0.1 | 0.4 | 0 | 0 | 0.1 |
| 9.5 | - | 40 | 35 | - | - | - | 0.1 | 52.7 | strong* | 54 | 53.4 |
| 9.5 | - | 40 | 20 | - | - | - | 0.1 | 4.5 | 61.6 | 48.3 | 38.1 |
| 9.5 | - | 40 | 35 | - | - | 5 | 0.1 | 45.4 | 35.6 | 7.9 | 29.6 |
| 9.5 | - | 10 | 20 | - | - | - | 0.1 | 32.7 | 23.5 | 20.4 | 25.5 |
| 9.5 | - | - | - | - | 1 | - | 0.1 | 17.8 | 45.4 | 8.2 | 23.8 |
| 9.5 | - | 40 | - | - | - | - | 0.1 | 4.6 | 2 | 12 | 6.2 |
| 9.5 | - | - | - | 3 | - | - | 0.1 | 13.8 | 0 | 0 | 4.6 |
| 9.5 | - | - | - | 3 | 1 | - | 0.1 | 0 | 6.2 | 6.5 | 4.2 |
| 9.5 | 20 | 25 | 25 | 1 | - | - | 0.1 | 3.2 | 2.6 | 0 | 1.9 |
| 9.5 | - | - | - | 3 | 1 | 5 | 0.1 | 0.1 | 2.8 | 0 | 1.0 |
| 9.5 | - | 10 | 35 | - | - | - | 0.1 | 1.2 | 1.3 | 0 | 0.8 |
| 9.5 | 70 | - | - | - | - | - | 10 | 2.1 | 0 | 0 | 0.7 |
| 9.5 | 45 | - | - | - | - | - | 4 | 0.4 | 0 | 0 | 0.1 |
| 9.5 | 70 | - | - | - | - | 5 | 10 | 0 | 0 | 0.3 | 0.1 |
| 9.5 | - | - | - | - | - | - | 0.1 | 0 | 0 | 0 | 0.0 |

*Ca = calcium; Mg = magnesium; Si = silicon; P = phosphorus; Zn = zinc; NOM = natural organic matter; HCO3- = bicarbonate*

*0.15 mM $SO_4^{2-}$ added to all tests*

* *specimen broke due to instrument malfunction, but appeared to be quite strong*

Strength gain comparisons were made between containers with similar water chemistries during the in-depth study. In this shorter term study, results were highly variable for specimens within the same container so ascertaining whether different water quality parameters caused a significant statistical difference was difficult. For example, consider a comparison between the pH 9.5 water containing only bicarbonate and the pH 9.5 water containing 40 mg/L Mg and 20 mg/L Si (Figure 5.1). The strength of the specimens in this test was not statistically significantly different from that of the specimens in the control water ($p = 0.079$). The visible precipitate in one specimen and the high strength gain in two of the three specimens in this test, however, add credence to the theory that magnesium and silicon play an important role in autogenous healing.

Other tests of water containing magnesium and silicon showed interesting trends as well. A comparison of pH 9.5 water containing 10 mg/L magnesium and varying amounts of silicon shows that an intermediate amount of silicon (20 mg/L) is more conducive to autogenous healing strength gain than waters with no silicon or 35 mg/L silicon (Figure 5.2). One possible explanation is the formation of an amorphous magnesium silicate precipitate (described in detail in Chapter 7). Comparisons between pH 8 water with 40 mg/L magnesium and various amounts of silicon show a similar trend as in the pH 9.5 waters (Figure 5.1). Little strength is developed in pH 8 waters with 10 mg/L magnesium and no dependence on silicon concentration was observed (Figure 5.2).

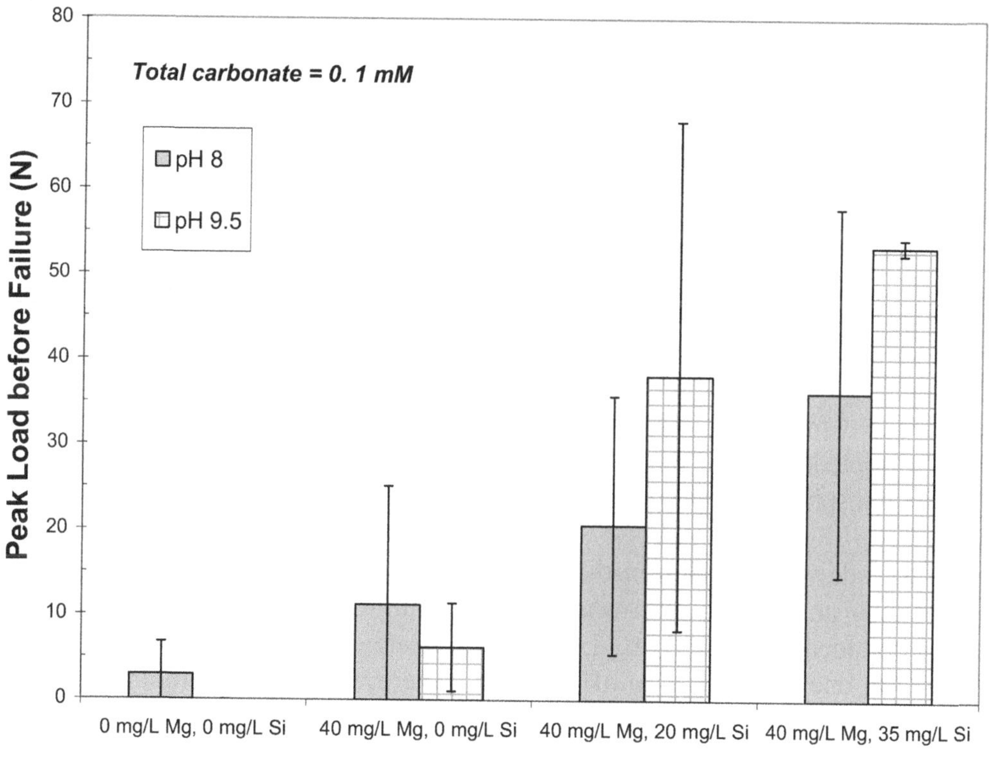


**Figure 5.1 Comparison of tests with 40 mg/L magnesium and varying amounts of silicon. Note increasing strength as amount of silicon is increased. (Error bars indicate ± 1 standard deviation.)**

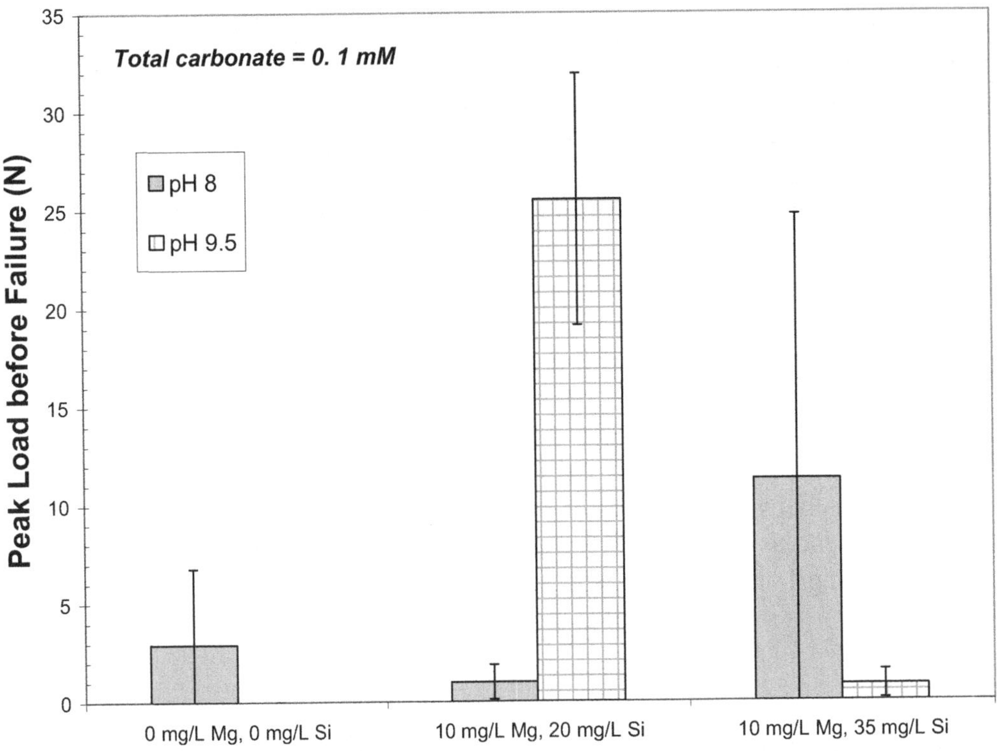


**Figure 5.2 Comparison of tests with 10 mg/L magnesium and varying amounts of silicon. Note optimum strength at intermediate amount of silicon. (Error bars indicate ± 1 standard deviation.)**

Orthophosphate addition (3 mg/L as P) did not affect the strength gain in contrast to what was observed in the screening study ($p = 0.42$). The addition of 1 mg/L zinc, on the other hand, showed signs of improving the strength gain ($p = 0.083$). In fact, the addition of 3 mg/L phosphorus coupled with 1 mg/L zinc actually led to less strength gain than the addition of zinc alone (Figure 5.3). Coupled with the data from the screening study, this seems to indicate an 'optimum' amount of phosphorus may be required and that overdosing may hinder autogenous healing.

Once again, the addition of calcium did not appear to enhance autogenous healing. In waters at pH 9.5 virtually no strength gain was observed with 45 or 70 mg/L calcium addition (these waters also contained 4 and 10 mg/L total carbonate, respectively). An interesting observation was made in one of the tests in pH 8 water. There was a minor amount of strength gain with 45 mg/L calcium addition, but none with 70 mg/L calcium addition. No explanation is readily apparent as calcium tends to leach into the water for most conditions tested.

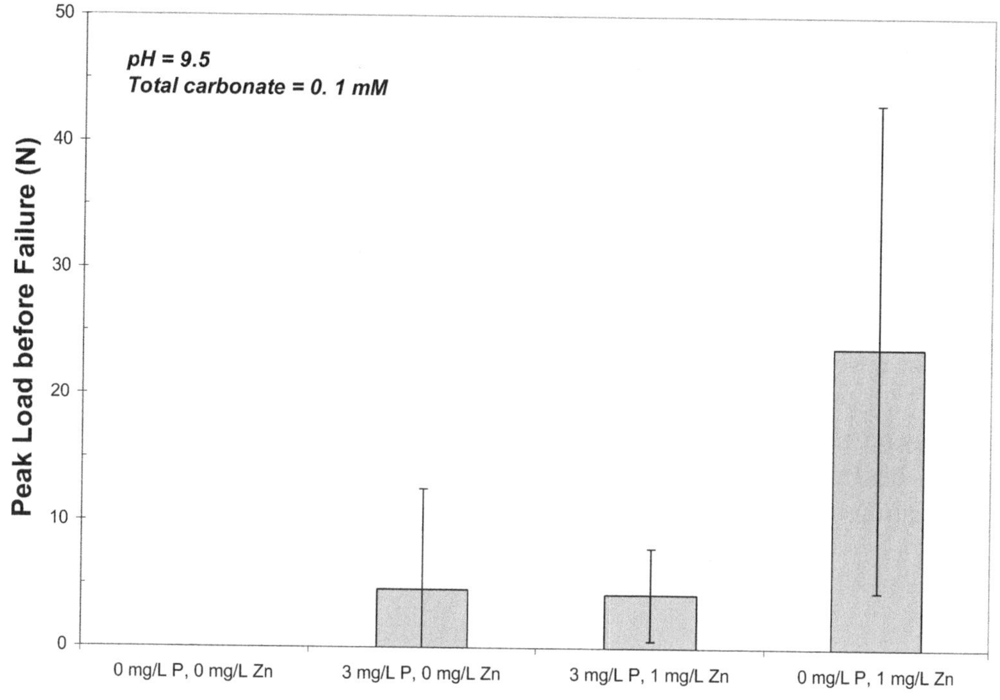


**Figure 5.3 Comparison of tests at pH 9.5 with varying amounts of orthophosphate and zinc. (Error bars indicate ± 1 standard deviation.)**

Three conditions were tested in which the only difference was 5 mg/L (as C) NOM addition. No significant differences were observed thus indicating NOM probably does not play a role in autogenous healing.

## CHLORIDE DIFFUSION TESTING

Several concrete specimens in which the 0.25 mm crack was 100% sealed (visually evaluated by recording fraction of crack through which light could be seen) were tested to determine the chloride diffusion rate through the 'healed' crack. A virgin 'uncracked' specimen was also evaluated for comparison. In no situation did an autogenously healed specimen approach the low rate of chloride diffusion through a virgin specimen (Figure 5.4). The virgin specimen was tested over a one month period and exhibited an average diffusion rate of $8.7 \times 10^{-6}$ mg $Cl^-$ / sec (Note: average diffusion rates determined from slope of 'best fit' line). Water leaked through many specimens immediately even though no light was visible through the 'healed' crack. The lowest amount of chloride diffusion was observed for a specimen from the pH 9.5 water with 40 mg/L magnesium and 20 mg/L silicon. This specimen also had a precipitate visible along the edge of the crack and was the strongest tested (62 N). The result of the chloride diffusion test was promising as only 1.3 mg $Cl^-$ / sec diffused through the 'healed' crack. While this rate of diffusion does not approach that of virgin concrete, it is orders of magnitude slower than an unhealed crack. Over time, as additional autogenous healing occurs, the rate of chloride diffusion is expected to decrease even more.

## CONCLUSIONS

- The presence of both magnesium and silicon is essential to forming strong 'glue' to seal concrete cracks.
- The addition of 3 mg/L phosphorus did not appear to enhance autogenous healing.
- The addition of 70 mg/L calcium (with 10 mM total carbonate) did not appear to enhance autogenous healing.
- The presence or absence of NOM did not affect autogenous healing.
- Chloride diffusion rates do not return to levels seen in virgin concrete when cracks are autogenously healed. However, chloride diffusion is sometimes impeded when autogenous healing occurs, regardless of the strength gain attained.

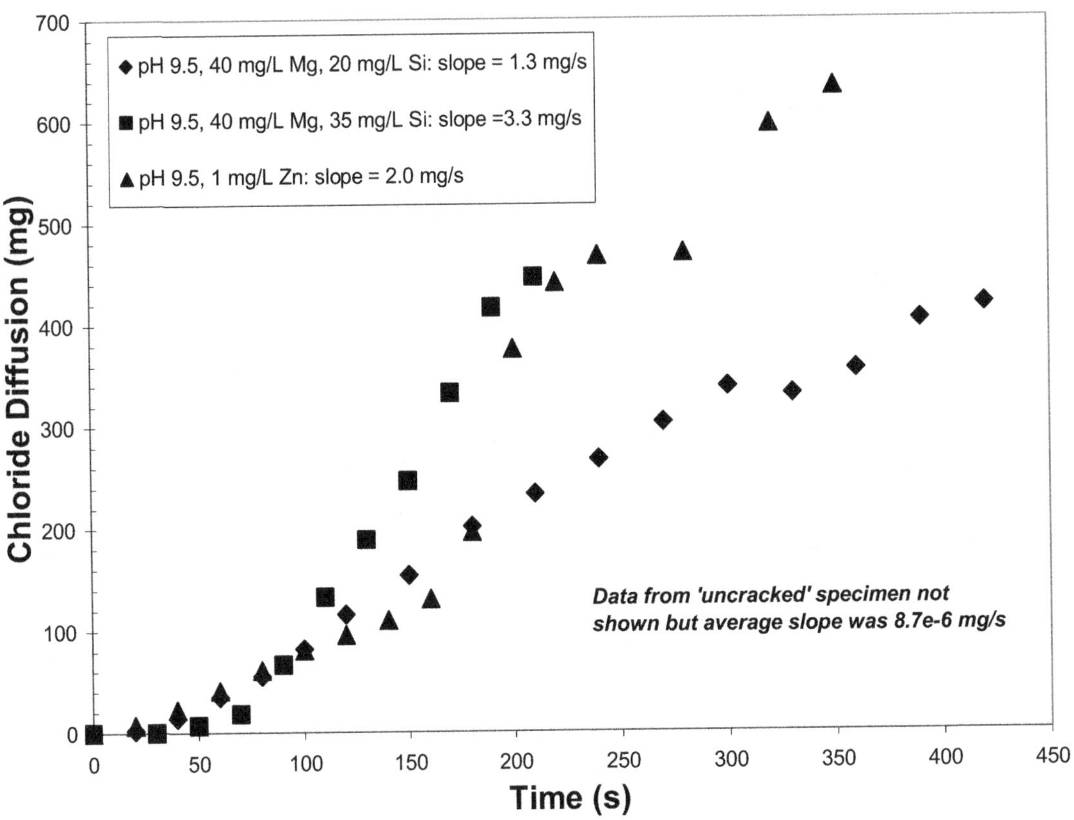


**Figure 5.4 Results from chloride diffusion experiment for selected specimens. Autogenously healed specimens are 5 orders of magnitude less effective at stopping chloride diffusion than a virgin concrete specimen.**

# CHAPTER 6: IDENTIFICATION OF PRECIPITATES RESPONSIBLE FOR AUTOGENOUSLY HEALED CRACKS

Visual evidence indicated that all water chemistries tested exhibited some autogenous healing. Strength testing and chloride diffusion testing results also indicate that some degree of healing is occurring (see Chapter 5). In this chapter we present results from an SEM-EDS study and a Raman spectroscopy study to assess what precipitates are forming on concrete surfaces that could be responsible for the autogenous healing phenomenon.

## SEM-EDS ANALYSIS

Two concrete specimens from the screening study were analyzed by SEM-EDS. One specimen from the pH 7 test with 25 mg/L Si had poor strength (41 N) and a second specimen from the pH 9.5 test with 25 mg/L Si, 25 mg/L Mg, 20 mg/L Ca, and 1 mg/L P had high strength (204 N). An EDS analysis of each surface indicated that magnesium and silicon were major components of the glue that formed within the cracks. While both surfaces were very heterogeneous, the surface of the 'poor strength' specimen had no deposits that contained magnesium (as expected since no Mg was added to the initial test water). Additionally, the only silicon observed on this surface appeared to be within the aggregate. The majority of the surface of the 'poor strength' specimen was most likely calcium oxide or calcium carbonate. The 'high strength' specimen, on the other hand, had a much rougher texture and contained fairly high levels of magnesium in some areas. Low levels of aluminum were also observed on this surface. While the surface was variable, the molar ratio of the magnesium silicate precipitate is 2.3 Si:Mg ($\pm$ 1.9). The high variability was most likely due to quartz particles in the aggregate and the precision of the instrument (the smallest area on which an EDS reading can be taken is 1 micrometer x 1 micrometer). An additional reason for high variability was that EDS readings were taken directly on the face of the concrete specimen, and therefore, could include elemental analyses of concrete constituents in addition to the precipitate.

All six specimens from the in-depth study that had a visible precipitate near the crack surface were tested with SEM-EDS. In this study, the precipitate was scraped off the crack surface and placed on an EDS mount prior to analysis. With this method we hoped to be able to focus on the 'glue' without interference from the concrete surface. Two specimens (pH 9.5, 40 mg/L Mg, 20 mg/L Si; pH 9.5, 40 mg/L Mg, 35 mg/L Si) had numerous deposits of a solid with an approximately equimolar ratio of Si:Mg (Table 6.1 and Figures 6.1, 6.2). These were also two of the strongest specimens (peak load before failure of the pH 9.5, 40 mg/L Mg, 35 mg/L Si specimen was 62 N).

Only calcium oxides and calcium carbonate were observed on the surface of the specimen from the water at pH 9.5 with 1 mg/L zinc. This specimen did exhibit a relatively large strength (45 N).

Numerous deposits of minerals with atomic composition consistent with apatite [$Ca_5(PO_4)_3OH$] were found on the surfaces of both specimens analyzed from the pH 9.5 water with 3 mg/L P and 1 mg/L Zn; however, the strength of these was very low (6 N each) (Figure 6.3).

**Table 6.1**

**EDS data for selected specimens from in-depth study**

| Test Condition | SEM-EDS Elemental Composition (%) | | | | | | | Possible Compound |
|---|---|---|---|---|---|---|---|---|
| | Ca | C* | O | Mg | Si | P | Zn | |
| pH 9.5, 40 mg/L Mg, 35 mg/L Si, 0.1 mM $HCO_3^-$ | 28.1 | | 70.9 | 0.4 | 0.5 | | | $Ca(OH)_2$ or $CaCO_3$ |
| | 14.5 | | 52.9 | 14.1 | 18.5 | | | Mg-silicate |
| | 18.7 | | 79.0 | 1.3 | 0.9 | | | $Ca(OH)_2$ or $CaCO_3$ |
| | 24.9 | | 74.4 | 0.2 | 0.5 | | | $Ca(OH)_2$ or $CaCO_3$ |
| | 0.7 | | 70.1 | 15.7 | 13.5 | | | Mg-silicate |
| | 19.7 | | 76.2 | 2.0 | 2.0 | | | $Ca(OH)_2$ or $CaCO_3$ |
| | 27.9 | | 71.6 | 0.2 | 0.3 | | | $Ca(OH)_2$ or $CaCO_3$ |
| | 19.0 | | 68.3 | 5.3 | 7.3 | | | $Ca(OH)_2$ or $CaCO_3$ |
| | 0.7 | | 65.4 | 16.3 | 17.6 | | | Mg-silicate |
| | 15.6 | | 68.0 | 7.0 | 9.3 | | | $Ca(OH)_2$ or $CaCO_3$ |
| pH 9.5, 40 mg/L Mg, 20 mg/L Si, 0.1 mM $HCO_3^-$ | 1.0 | | 70.3 | 16.7 | 12.0 | | | Mg-silicate |
| | 16.4 | 1.2 | 81.7 | 0.5 | 0.3 | | | $Ca(OH)_2$ or $CaCO_3$ |
| | 6.0 | | 69.6 | 13.4 | 10.9 | | | Mg-silicate |
| | 0.1 | | 70.6 | 15.2 | 13.9 | | | Mg-silicate |
| | 5.3 | | 74.3 | 10.7 | 9.6 | | | Mg-silicate |
| pH 9.5, 1 mg/L Zn, 0.1 mM $HCO_3^-$ | 21.2 | | 78.5 | | | | | $Ca(OH)_2$ or $CaCO_3$ |
| | 18.5 | 22.0 | 59.3 | | | | | $Ca(OH)_2$ or $CaCO_3$ |
| | 20.8 | | 78.9 | | | | | $Ca(OH)_2$ or $CaCO_3$ |
| | 48.2 | | 51.2 | | | | 0.4 | $Ca(OH)_2$ or $CaCO_3$ |
| | 22.3 | | 77.3 | | | | | $Ca(OH)_2$ or $CaCO_3$ |
| pH 9.5, 1 mg/L Zn, 3 mg/L P, 0.1 mM $HCO_3^-$ | 13.7 | | 77.6 | | 0.3 | 8.3 | | $Ca_5(PO_4)_3(OH)$ |
| | 17.7 | | 71.0 | | 0.3 | 10.9 | | $Ca_5(PO_4)_3(OH)$ |
| | 38.6 | | 43.5 | | 0.6 | 16.8 | 0.4 | ? |
| | 16.3 | | 73.2 | | 0.4 | 10.0 | | $Ca_5(PO_4)_3(OH)$ |
| | 15.6 | | 74.4 | | 0.3 | 9.6 | | $Ca_5(PO_4)_3(OH)$ |
| | 8.6 | | 76.8 | | 0.6 | 13.6 | 0.3 | ? |
| | 5.6 | 15.7 | 69.4 | | 0.4 | 8.8 | | ? |
| | 7.7 | | 79.0 | | 1.4 | 11.9 | | ? |
| | 10.1 | | 73.8 | | 0.3 | 15.3 | 0.4 | ? |
| | 16.0 | | 28.0 | | 0.5 | 49.8 | 5.6 | ? |

**Instrument malfunctioned periodically when analyzing for carbon; no measurement for carbon does not necessarily mean carbon was not present.*

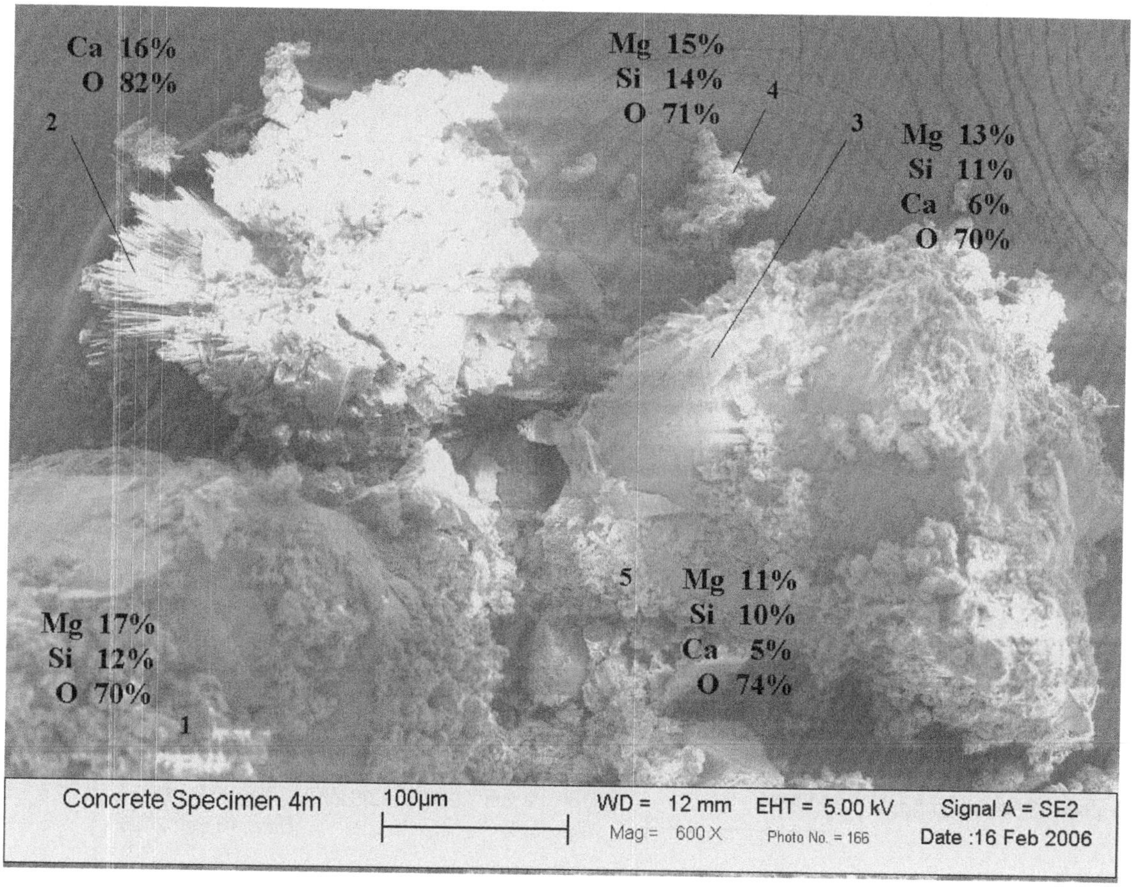


**Figure 6.1 SEM micrograph of precipitate scraped from concrete surface. Specimen was submerged in pH 9.5 water with 40 mg/L Mg, 20 mg/L Si, and 0.1 mM carbonate. Elemental composition in molar %.**

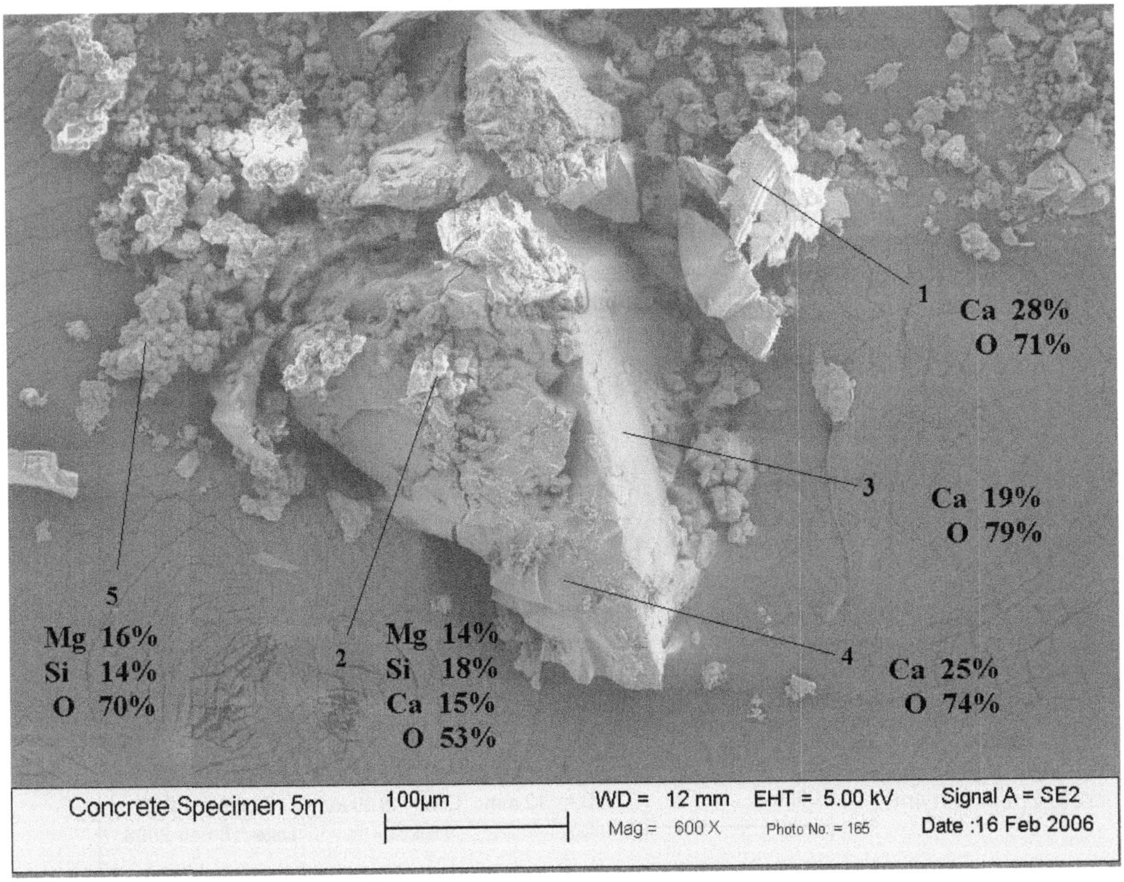


**Figure 6.2 SEM micrograph of precipitate scraped from concrete surface. Specimen was submerged in pH 9.5 water with 40 mg/L Mg, 35 mg/L Si, and 0.1 mM carbonate. Elemental composition in molar %.**

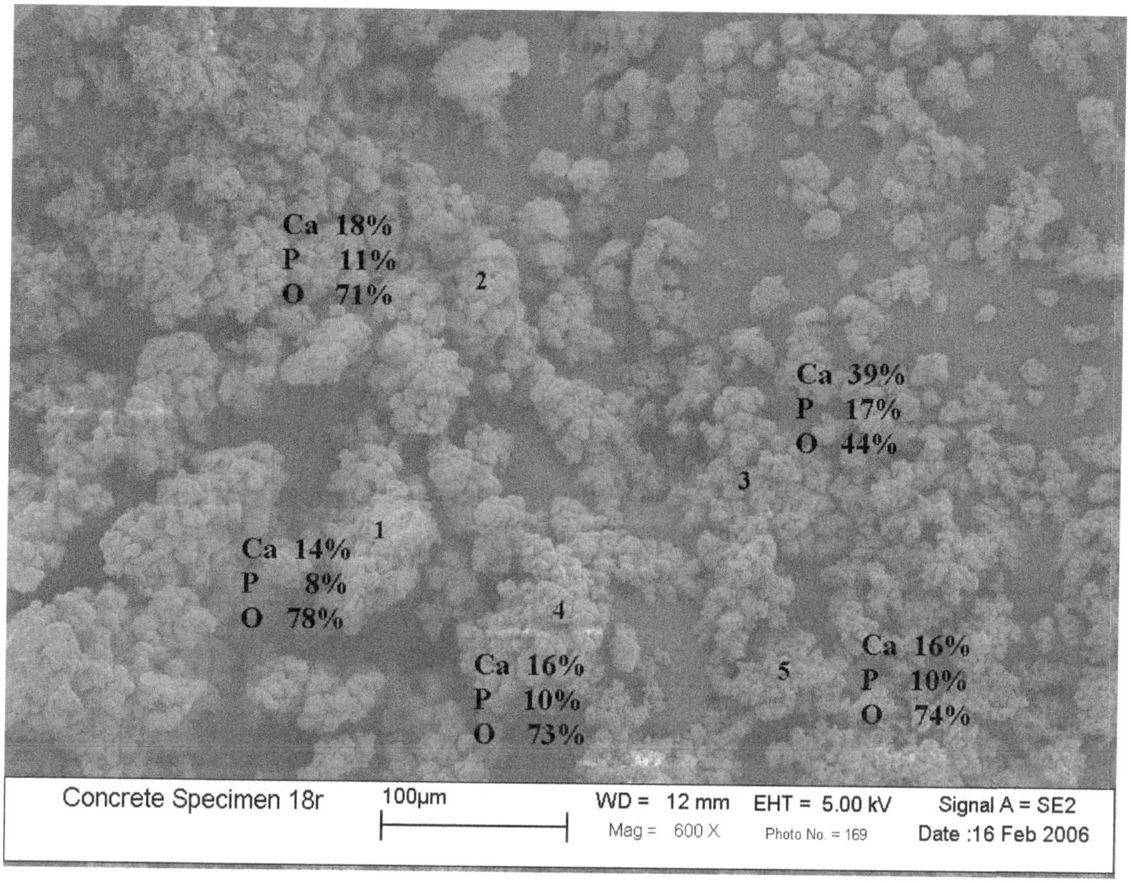


**Figure 6.3 SEM micrograph of precipitate scraped from concrete surface. Specimen was submerged in pH 9.5 water with 3 mg/L P, 1 mg/L Zn, and 0.1 mM carbonate. Elemental composition in molar %.**

## RAMAN SPECTROSCOPY ANALYSIS

Several specimens from the in-depth study, including all specimens which had a visible precipitate near the crack surface, were analyzed with Raman spectroscopy. Comparisons between spectral peaks produced by Raman analysis of the coupons (Table 6.2) and peaks found in minerals that are likely present in the 'glue' (Table 6.3) made it evident that carbonate minerals were present. However, two peaks (one at 888 $cm^{-1}$ and one at 437 $cm^{-1}$) were present in several of the stronger specimens and could not be identified using existing databases. Since the importance of magnesium and silicon to the development of strong 'glue' had been established in each study, magnesium silicate minerals were included in the search. Although no match was found, past research by the project team has detected an amorphous magnesium silicate precipitate with a stoichiometry similar to that of the mineral forsterite ($Mg_2SiO_4$; Parks and Edwards, 2007). In that study amorphous $Mg_2SiO_4$ precipitated from solutions containing various quantities of magnesium and silicon when pH exceeded 10.4. A sample of this precipitate was formed using procedures described by Parks and Edwards (2007) and then analyzed with Raman spectroscopy (Table 6.3).

A comparison of the spectra indicates that aragonite, magnesium-calcite, and/or the amorphous magnesium silicate comprise the 'glue' in most specimens (Table 6.4). All high strength specimens tested had the amorphous magnesium silicate present.

These analyses support the conclusion that the formation of an amorphous magnesium silicate seems to increase the overall glue strength. Additionally, phosphorous and zinc do not appear to become incorporated within the glue.

## COMPARISON OF RESULTS

In general there was very good agreement between SEM-EDS and Raman spectroscopy results. Both techniques identified magnesium silicate, calcium carbonate (aragaonite), and silicon dioxide (coesite or quartz) as compounds present on the surfaces of autogenously healed concrete. SEM-EDS identified a compound stoichiometrically similar to apatite on the surfaces of concrete that had been exposed to phosphate; however, this compound went undetected by Raman analysis.

**Table 6.2**
**Spectral peaks for specimens tested by Raman spectroscopy**

| Initial Water Condition | Peak Load (N) | Raman shift* ($cm^{-1}$) | | | | | | | | | | | | | | | | | | | |
|---|---|---|---|---|---|---|---|---|---|---|---|---|---|---|---|---|---|---|---|---|---|
| pH 9.5, 10 mg/L Mg, 20 mg/L Si | 32.7 | <u>154</u> | | 206 | | | | | | 437 | | | | | 706 | | | 945 | <u>1087</u> | | |
| pH 9.5, 40 mg/L Mg, 20 mg/L Si | 61.6 | <u>154</u> | | 208 | | | | 341 | | <u>437</u> | | 488 | | | | 788 | 888 | | <u>1086</u> | | |
| pH 9.5, 40 mg/L Mg, 35 mg/L Si | 52.7 | <u>154</u> | | | | | | | | <u>437</u> | | 488 | 524 | | | | <u>888</u> | | | 1104 | |
| pH 8, 10 mg/L Mg, 35 mg/L Si | 26.6 | | 178 | | | | | 340 | 381 | <u>437</u> | | 488 | | 651 | | | <u>888</u> | | <u>1087</u> | | 1203 |
| pH 8, 40 mg/L Mg, 20 mg/L Si | 37.8 | | | | | | | | | 437 | <u>466</u> | 488 | | | | | | | | | |
| pH 8, 40 mg/L Mg, 35 mg/L Si | 49.6 | | | | | | | | | <u>437</u> | | 488 | | | | | 888 | | 1086 | | |
| pH 9.5, 3 mg/L P, 1 mg/L Zn | 6.5 | 154 | | | <u>282</u> | | | | | | | | | | 714 | | | | <u>1087</u> | | |
| pH 9.5, 1 mg/L Zn | 45.4 | <u>154</u> | | | <u>283</u> | | | | | 437 | | | | | <u>713</u> | | 888 | | <u>1087</u> | | |
| pH 9.5, 40 mg/L Mg, 35 mg/L Si, 5 mg/L NOM | 45.4 | | 178 | | | 300 | 322 | <u>340</u> | | <u>437</u> | | 488 | | 651 | | | <u>888</u> | | | 1104 | 1204 |

** Major peaks are underlined*

**Table 6.3**
**Raman peaks for typical minerals found in concrete and various magnesium silicate minerals**

| Mineral | Raman shift* ($cm^{-1}$) | Formula |
|---|---|---|
| andradite | **350** 369 **874** | Ca3Fe2(SiO4)3 |
| apatite | 592 963 1069 | Ca5(PO4)3(OH) |
| aragonite | 154 **205** 704 **1085** | CaCO3 |
| biotite | 178 352 **983** | K(Fe,Mg)3AlSi3O10(OH)2 |
| calcite | **281** 712 **1088** | CaCO3 |
| calcium aluminate | 506 **754** | Ca3Al2O6 |
| CH | 254 356 **677** | Ca(OH)2 |
| chrysotile | 436 625 695 | Mg3Si2O5(OH)4 |
| coesite | 155 210 **273** **467** **520** 839 1145 | SiO2 |
| C-S-H Gel | **450** **620** 680 **850** | - |
| dicalcium silicate | 422 516 536 **857** 976 | Ca2SiO4 |
| dolomite | 175 298 712 1097 | CaMg(CO3)2 |
| enstatite | 235 342 401 662 684 1011 | MgSiO3 |
| ettringite | 450 615 **988** | Ca6Al2(SO4)3(OH)12.26H2O |
| forsterite | 304 440 609 821 853 963 1114 | Mg2SiO4 |
| galena | 269 359 | PbS |
| gypsum | 491 **626** **1010** 1135 | CaSO4.2H2O |
| hematite | **223** **289** 412 501 615 1296 | Fe2O3 |
| iron calcium aluminate | **256** 308 737 | Ca4Fe2Al2O10 |
| kyanite | 301 **487** 951 | Al2SiO5 |
| magnesium-calcite | 155 **281** 712 **1087** | CaMg(CO3)2 |
| quartz | **206** 355 **468** **1089** | SiO2 |
| siderite | **289** **300** 732 **1090** | FeCO3 |
| smithsonite | 193 303 1093 | ZnCO3 |
| talc | 193 361 676 | Mg3Si4O10(OH)2 |
| thaumasite | 418 454 658 **990** **1071** | Ca3Si(CO3)(SO4)(OH)6.12H2O |
| thenardite | 994 1103 1153 | Na2SO4 |
| tricalcium silicate | 517 544 **838** | Ca3SiO5 |
| vaterite | **1075** 1081 **1090** | CaCO3 |
| am-magnesium silicate† | 178 322 340 381 **437** 488 **888** | |

** Major peaks are underlined*

*†amorphous magnesium silicate formed in jar test conducted at pH 10.8 and containing 48 mg/L Mg and 12 mg/L Si*

**Table 6.4**
**Comparison of spectral peaks for specimens tested by Raman spectroscopy**

| **Initial Water Condition /Mineral** | **Raman shift* (cm$^{-1}$)** | | | | | | | | | | | | | | | | | | | |
|---|---|---|---|---|---|---|---|---|---|---|---|---|---|---|---|---|---|---|---|---|
| pH 9.5, 10 mg/L Mg, 20 mg/L Si | $\underline{154}$ | | 206 | | | | | | 437 | | | | | 706 | | | 945 | $\underline{1087}$ | | |
| aragonite | 154 | | $\underline{205}$ | | | | | | | | | | | 704 | | | | $\underline{1085}$ | | |
| pH 9.5, 40 mg/L Mg, 20 mg/L Si | $\underline{154}$ | | 208 | | | | 341 | | $\underline{437}$ | | 488 | | | | 788 | 888 | | $\underline{1086}$ | | |
| aragonite | 154 | | $\underline{205}$ | | | | | | | | | | | 704 | | | | $\underline{1085}$ | | |
| am-magnesium silicate | | 178 | | | | 322 | 340 | 381 | $\underline{437}$ | | 488 | | | | | $\underline{888}$ | | | | |
| pH 9.5, 40 mg/L Mg, 35 mg/L Si | $\underline{154}$ | | | | | | | | $\underline{437}$ | | 488 | 524 | | | | $\underline{888}$ | | | 1104 | |
| am-magnesium silicate | | 178 | | | | 322 | 340 | 381 | $\underline{437}$ | | 488 | | | | | $\underline{888}$ | | | | |
| pH 8, 10 mg/L Mg, 35 mg/L Si | | 178 | | | | | 340 | 381 | $\underline{437}$ | | 488 | | 651 | | | $\underline{888}$ | | $\underline{1087}$ | | 1203 |
| am-magnesium silicate | | 178 | | | | 322 | 340 | 381 | $\underline{437}$ | | 488 | | | | | $\underline{888}$ | | | | |
| pH 8, 40 mg/L Mg, 20 mg/L Si | | | | | | | | | 437 | $\underline{466}$ | 488 | | | | | | | | | |
| am-magnesium silicate | | 178 | | | | 322 | 340 | 381 | $\underline{437}$ | | 488 | | | | | $\underline{888}$ | | | | |
| coesite | 155 | | 210 | $\underline{273}$ | | | | | | $\underline{467}$ | | $\underline{520}$ | | | 839 | | | | 1145 | |
| pH 8, 40 mg/L Mg, 35 mg/L Si | | | | | | | | | $\underline{437}$ | | 488 | | | | | 888 | | 1086 | | |
| am-magnesium silicate | | 178 | | | | 322 | 340 | 381 | $\underline{437}$ | | 488 | | | | | $\underline{888}$ | | | | |
| pH 9.5, 3 mg/L P, 1 mg/L Zn | 154 | | | $\underline{282}$ | | | | | | | | | | 714 | | | | $\underline{1087}$ | | |
| magnesium-calcite | 155 | | | $\underline{281}$ | | | | | | | | | | 712 | | | | $\underline{1087}$ | | |
| pH 9.5, 1 mg/L Zn | $\underline{154}$ | | | $\underline{283}$ | | | | | 437 | | | | | $\underline{713}$ | | 888 | | $\underline{1087}$ | | |
| magnesium-calcite | 155 | | | $\underline{281}$ | | | | | | | | | | 712 | | | | $\underline{1087}$ | | |
| am-magnesium silicate | | 178 | | | | 322 | 340 | 381 | $\underline{437}$ | | 488 | | | | | $\underline{888}$ | | | | |
| pH 9.5, 40 mg/L Mg, 35 mg/L Si, 5 mg/L NOM | | 178 | | | 300 | 322 | $\underline{340}$ | | $\underline{437}$ | | 488 | | 651 | | | $\underline{888}$ | | | 1104 | 1204 |
| am-magnesium silicate | | 178 | | | | 322 | 340 | 381 | $\underline{437}$ | | 488 | | | | | $\underline{888}$ | | | | |

** Major peaks are underlined*

# CHAPTER 7: SUMMARY, CONCLUSIONS, AND RECOMMENDATIONS

Autogenous healing is the phenomenon that occurs when hairline cracks that develop in concrete repair themselves through reactions with constituents in water. An understanding of when and how this healing occurs as opposed to cases where these cracks enlarge and contribute to concrete failure would be beneficial to water utilities.

This research project examined autogenous healing in a variety of experimental waters. The water quality parameters were selected to span the range of key ion concentrations that are encountered in practice and to focus primarily on elements that may play a role in the autogenous healing process. Concrete coupons with a fixed crack width of 0.25 mm were utilized and tests were performed in triplicate. Waters with submerged coupons were gently agitated for 7 to 12 months. Throughout each test, water was periodically poured off and replaced with new. Water samples were withdrawn and analyzed for various metal and anion concentrations. At the end of the testing period, concrete specimens were evaluated for autogenous healing by several criteria. Select specimens were tested for their ability to hinder chloride diffusion. All specimens were subjected to a strength test. Finally, the interior surfaces of select specimens were evaluated for carbonation. SEM-EDS and Raman spectroscopy were used in an attempt to identify precipitates that may have formed on concrete surfaces.

Some degree of autogenous healing (as evidenced by the absence of light passing through the simulated crack) occurred in each of the water conditions tested. There appeared to be two types of healing, however. Healing resulted in substantial strength development in some specimens – primarily in pH 9.5 water with high levels of magnesium and silicon. In other specimens, healing occurred with no appreciable gain in strength. Although high calcium concentrations are touted as inhibiting concrete corrosion (e.g. LI) and calcium carbonate is cited as being the 'glue' responsible for autogenous healing, we did not see significant strength gains in this study when calcium was added to test waters.

## CONCLUSIONS

Conclusions that can be taken from this work include:

- Autogenous healing can be an effective method of sealing small cracks that may develop in concrete.
- Even though a high strength seal can be developed, the formation of a precipitate on concrete surfaces at pH 9.5 when high levels of magnesium and silicon are present does not inhibit calcium leaching. On the contrary, soluble calcium increases when high levels of magnesium and silicon are present. This result emphasizes the conclusion that LI is not a viable tool for predicting autogenous healing.
- Carbonation of internal concrete surfaces was not detected when cracks were sealed via autogenous healing.
- Chloride diffusion rates do not return to levels seen in virgin concrete when cracks are autogenously healed. However, water permeability and chloride diffusion are impeded in some cases, regardless of the strength attained. The least amount of chloride diffusion was observed for a specimen from the pH 9.5 water with 40 mg/L magnesium and 20

mg/L silicon. The result of the chloride diffusion test was promising as only 1.3 mg $Cl^-$ / sec diffused through the 'healed' crack.

## RECOMMENDATIONS

Additional research needs include more tests using hydraulic conditions that would be typical in a drinking water distribution system. We anticipate that autogenous healing would still occur but possibly at a much slower rate so longer term studies should be conducted. Better analytical techniques need to be evaluated (such as X-ray diffraction) which will allow positive identification of the 'glue' material.

Strength data for shorter exposure periods could provide additional insight into the autogenous healing process. White precipitates on crack surfaces were visible after as little as 6 weeks of exposure in some situations. A follow-up study in which strength tests could be conducted on sacrificial coupons at 30 day intervals would be invaluable. One caveat, however, is that a high level of variability was observed in these tests and therefore, results from single sacrificial coupons may not be indicative of the degree of autogenous healing of other test coupons subjected to the same water quality parameters.

While similar trends were observed in both series of tests, the four-fold higher strength gain observed in the 12 month study as compared to the subsequent 7 month study is a concern. While varying solution chemistries may account for some of this inconsistency, the authors feel that the test conditions greatly affected this outcome. Large stagnation intervals and limited pH control in the 12 month study would be expected to be conducive for the occurrence of autogenous healing. That is, during periods of stagnation pore water pH could increase dramatically which could then lead to higher levels of precipitates responsible for the autogenous healing phenomenon. Future research, as stated above, should make a distinction between stationary and flowing water.

The authors believe various factors attributed to the high degree of variability exhibited by replicate coupons. A major factor appeared to be the heterogeneity of the concrete specimens. Coupons were a fairly small size and the presence or absence of large aggregates could affect the coupons response to various water quality parameters. Future research should perhaps eliminate the large aggregate fraction and utilize only cement, sand, and water to cast concrete coupons.

Concrete age could be another factor that might enhance or impede autogenous repair. Coupons used in this study were 3-5 years old. Older concrete, such as that found in many distribution systems, may behave differently.

The use of the Langelier Index to determine whether concrete corrosion will occur in distribution systems is widely accepted although it is not always reliable. In this study we have observed that LI cannot indicate when appreciable autogenous healing of cracks in concrete will occur with restoration of strength. Obviously, other factors such as magnesium and silicon concentrations in the treated water play an important role and must be explicitly considered if an index were to be developed that describes autogenous repair.

For utilities with low amounts of magnesium and silicon in their treated water, we speculate that high rates of concrete degradation could occur due to the absence of autogenous healing. Small cracks will tend to enlarge over time without autogenous healing eventually leading to water leaks in the distribution system. A concrete liner or seal coat might be the best defense against aggressive water (i.e., water with low levels of magnesium and silicon) in this instance. While it may be possible to promote healing by manipulating the environment (water chemistry) or the concrete composition, the interplay between these two factors needs to be further evaluated.

# APPENDIX A: MISCELLANEOUS RAW DATA

**Table A.1**

**Timetable for in-depth study**

| Date | Day # | Comment | ICP-MS Samples | IC Samples | TOC Samples | Conductance Measurements | Visual Observations |
|---|---|---|---|---|---|---|---|
| 6/2/2005 | 0 | **Begin experiment** | yes | | | yes | |
| 6/4/2005 | 2 | | yes | | | | |
| 6/8/2005 | 6 | | yes | yes | | | |
| 6/9/2005 | 7 | Water change | yes | | | | |
| 6/15/2005 | 13 | | | | | yes | |
| 6/16/2005 | 14 | Water change | yes | | | | |
| 7/14/2005 | 42 | Water change | yes | | | | yes |
| 8/11/2005 | 70 | Water change | yes | | yes | | yes |
| 9/9/2005 | 99 | | | yes | | yes | |
| 9/12/2005 | 102 | Water change | yes | | | | yes |
| 10/20/2005 | 140 | Water change | yes | | | | yes |
| 11/8/2005 | 159 | | | | | yes | |
| 11/14/2005 | 165 | | | | yes | | |
| 11/15/2005 | 166 | | | yes | | | |
| 11/29/2005 | 180 | Water change | yes | | | | yes |
| 1/12/2006 | 224 | | | | | | yes |
| 1/17/2006 | 229 | Chloride diffusion testing | | | | | |
| 1/20/2006 | 232 | **End experiment** | yes | | | | yes |
| 1/23/2006 | 235 | Strength testing | | | | | |
| 1/25/2006 | 237 | Carbonation testing | | | | | |

Conductance (Table A.2), IC (Table A.3), TOC (Table A.4), and ICP-MS (Table A.5) measurements are included as part of Appendix A.

Digital images of selected specimens are included in Appendices B and C.

**Table A.2**
**Conductance measurements during in-depth study**

| Test # | pH | Ca mg/L | Mg mg/L | Si mg/L | P mg/L | Zn mg/L | NOM mg C/L | $HCO_3^-$ mM | Est. Conductance* (μmho) | Conductance (μmho) 06/02/05 | 06/15/05 | 09/09/05 | 11/08/05 |
|---|---|---|---|---|---|---|---|---|---|---|---|---|---|
| 1 | 9.5 | - | - | - | - | - | - | 0.1 | 0.03 | 0.07 | 0.12 | 0.15 | 0.14 |
| 2 | 9.5 | - | 10 | 20 | - | - | - | 0.1 | 0.16 | 0.24 | 0.33 | 0.33 | 0.32 |
| 3 | 9.5 | - | 10 | 35 | - | - | - | 0.1 | 0.19 | 0.32 | 0.41 | 0.45 | 0.40 |
| 4 | 9.5 | - | 40 | 20 | - | - | - | 0.1 | 0.39 | 0.39 | 0.50 | 0.54 | 0.56 |
| 5 | 9.5 | - | 40 | 35 | - | - | - | 0.1 | 0.42 | 0.47 | 0.60 | 0.63 | 0.63 |
| 6 | 9.5 | - | 40 | - | - | - | - | 0.1 | 0.34 | 0.30 | 0.39 | 0.43 | 0.41 |
| 7 | 8 | - | - | - | - | - | - | 0.1 | 0.03 | 0.06 | 0.15 | 0.15 | 0.15 |
| 8 | 8 | - | 10 | 20 | - | - | - | 0.1 | 0.16 | 0.25 | 0.41 | 0.38 | 0.36 |
| 9 | 8 | - | 10 | 35 | - | - | - | 0.1 | 0.19 | 0.34 | 0.48 | 0.50 | 0.48 |
| 10 | 8 | - | 40 | 20 | - | - | - | 0.1 | 0.39 | 0.40 | 0.59 | 0.60 | 0.56 |
| 11 | 8 | - | 40 | 35 | - | - | - | 0.1 | 0.42 | 0.48 | 0.66 | 0.72 | 0.66 |
| 12 | 8 | - | 40 | - | - | - | - | 0.1 | 0.34 | 0.29 | 0.44 | 0.46 | 0.45 |
| 13 | 9.5 | 45 | - | - | - | - | - | 4 | 0.49 | 0.52 | 0.51 | 0.55 | 0.57 |
| 14 | 9.5 | 70 | - | - | - | - | - | 10 | 0.98 | 0.76 | 1.10 | 1.16 | 1.17 |
| 15 | 8 | 45 | - | - | - | - | - | 4 | 0.49 | 0.44 | 0.56 | 0.44 | 0.55 |
| 16 | 8 | 70 | - | - | - | - | - | 10 | 0.98 | 0.65 | 0.95 | 0.94 | 0.88 |
| 17 | 9.5 | - | - | - | 3 | - | - | 0.1 | 0.05 | 0.07 | 0.12 | 0.17 | 0.16 |
| 18 | 9.5 | - | - | - | 3 | 1 | - | 0.1 | 0.05 | 0.07 | 0.10 | 0.15 | 0.16 |
| 19 | 9.5 | - | - | - | - | 1 | - | 0.1 | 0.04 | 0.06 | 0.14 | 0.13 | 0.11 |
| 20 | 9.5 | - | 40 | 35 | - | - | 5 | 0.1 | 0.42 | 0.47 | 0.61 | 0.73 | 0.64 |
| 21 | 9.5 | 70 | - | - | - | - | 5 | 10 | 0.98 | 0.69 | 1.12 | 1.15 | 1.05 |
| 22 | 9.5 | - | - | - | 3 | 1 | 5 | 0.1 | 0.05 | 0.10 | 0.19 | 0.19 | 0.17 |
| 23 | 7 | - | - | - | - | - | - | 0.1 | 0.03 | 0.05 | 0.16 | 0.19 | 0.15 |
| 24 | 7 | 20 | 25 | 25 | 1 | - | - | 0.1 | 0.38 | 0.43 | 0.64 | 0.66 | 0.63 |
| 25 | 8 | 20 | 25 | 25 | 1 | - | - | 0.1 | 0.38 | 0.43 | 0.58 | 0.58 | 0.56 |
| 26 | 9.5 | 20 | 25 | 25 | 1 | - | - | 0.1 | 0.38 | 0.43 | 0.53 | 0.55 | 0.55 |

**Estimated conductance calculated from ionic strength of initial solution using equation derived by Russell (1976):*
*ionic strength = 0.000016 x conductance (μmho/cm)*

Table A.3

Chloride and sulfate measurements during in-depth study

| Test # | pH | Ca | Mg | Si | P | Zn | NOM | $HCO_3^-$ | $Cl^-$ (mg/L) | | | $SO_4^{2-}$ (mg/l) | | |
|---|---|---|---|---|---|---|---|---|---|---|---|---|---|---|
| | | mg/L | mg/L | mg/L | mg/L | mg/L | mg C/L | mM | 06/08/05 | 09/09/05 | 11/15/05 | 06/08/05 | 09/09/05 | 11/15/05 |
| 1 | 9.5 | - | - | - | - | - | - | 0.1 | 22.1 | 3.7 | 4.0 | 19.4 | 15.9 | 14.3 |
| 2 | 9.5 | - | 10 | 20 | - | - | - | 0.1 | 86.9 | 58.1 | 55.0 | 19.3 | 15.2 | 14.2 |
| 3 | 9.5 | - | 10 | 35 | - | - | - | 0.1 | 110.3 | 92.5 | 79.6 | 22.0 | 16.7 | 14.1 |
| 4 | 9.5 | - | 40 | 20 | - | - | - | 0.1 | 154.7 | 144.3 | 136.6 | 20.2 | 13.8 | 20.3 |
| 5 | 9.5 | - | 40 | 35 | - | - | - | 0.1 | 200.0 | 180.1 | 164.6 | 18.2 | 15.7 | 21.5 |
| 6 | 9.5 | - | 40 | - | - | - | - | 0.1 | 124.4 | 110.9 | 101.6 | 19.5 | 16.0 | 15.1 |
| 7 | 8 | - | - | - | - | - | - | 0.1 | 143.1 | 24.7 | 24.3 | 21.7 | 15.6 | 14.4 |
| 8 | 8 | - | 10 | 20 | - | - | - | 0.1 | 242.9 | 100.4 | 89.1 | 20.2 | 16.5 | 15.1 |
| 9 | 8 | - | 10 | 35 | - | - | - | 0.1 | 196.8 | 142.9 | 135.0 | 20.2 | 16.5 | 15.1 |
| 10 | 8 | - | 40 | 20 | - | - | - | 0.1 | 171.3 | 191.2 | 162.0 | 22.7 | 15.5 | 14.6 |
| 11 | 8 | - | 40 | 35 | - | - | - | 0.1 | 125.1 | 226.2 | 198.7 | 20.6 | 15.7 | 14.8 |
| 12 | 8 | - | 40 | - | - | - | - | 0.1 | 67.1 | 143.1 | 142.9 | 18.0 | 15.1 | 14.5 |
| 13 | 9.5 | 45 | - | - | - | - | - | 4 | 81.1 | 83.4 | 78.2 | 19.6 | 14.9 | 13.5 |
| 14 | 9.5 | 70 | - | - | - | - | - | 10 | 135.6 | 128.8 | 122.8 | 17.9 | 14.6 | 14.4 |
| 15 | 8 | 45 | - | - | - | - | - | 4 | 111.3 | 86.5 | 91.3 | 17.4 | 14.8 | 13.6 |
| 16 | 8 | 70 | - | - | - | - | - | 10 | 138.4 | 183.7 | 159.0 | 16.6 | 15.1 | 14.4 |
| 17 | 9.5 | - | - | - | 3 | - | - | 0.1 | 4.8 | 4.0 | 4.2 | 19.5 | 15.5 | 19.3 |
| 18 | 9.5 | - | - | - | 3 | 1 | - | 0.1 | 9.1 | 4.2 | 4.6 | 19.1 | 15.5 | 16.0 |
| 19 | 9.5 | - | - | - | - | 1 | - | 0.1 | 43.9 | 4.3 | 4.4 | 19.5 | 15.5 | 14.7 |
| 20 | 9.5 | - | 40 | 35 | - | - | 5 | 0.1 | 201.1 | 191.9 | 169.5 | 20.1 | 17.8 | 15.7 |
| 21 | 9.5 | 70 | - | - | - | - | 5 | 10 | 143.0 | 141.6 | 133.9 | 22.7 | 17.5 | 15.4 |
| 22 | 9.5 | - | - | - | 3 | 1 | 5 | 0.1 | 21.7 | 10.4 | 10.9 | 20.2 | 17.3 | 15.5 |
| 23 | 7 | - | - | - | - | - | - | 0.1 | 36.6 | 44.3 | 36.6 | 22.4 | 15.9 | 14.1 |
| 24 | 7 | 20 | 25 | 25 | 1 | - | - | 0.1 | 230.7 | 226.7 | 207.0 | 19.2 | 15.5 | 14.2 |
| 25 | 8 | 20 | 25 | 25 | 1 | - | - | 0.1 | 207.5 | 173.4 | 160.1 | 20.9 | 15.8 | 14.2 |
| 26 | 9.5 | 20 | 25 | 25 | 1 | - | - | 0.1 | 178.2 | 150.3 | 140.2 | 20.3 | 15.9 | 15.7 |

**Table A.4**
**Total carbon measurements during in-depth study**

| Test # | pH | Ca mg/L | Mg mg/L | Si mg/L | P mg/L | Zn mg/L | NOM mg C/L | $HCO_3^-$ mM | Calculated TC (mg/L as C) | August 11 Measurements TOC (mg/L as C) | August 11 Measurements TIC (mg/L as C) | August 11 Measurements TC (mg/L as C) | November 14 Measurements TOC (mg/L as C) | November 14 Measurements TIC (mg/L as C) | November 14 Measurements TC (mg/L as C) |
|---|---|---|---|---|---|---|---|---|---|---|---|---|---|---|---|
| 1 | 9.5 | - | - | - | - | - | - | 0.1 | 1.2 | 0.2 | 10.9 | 11.1 | 0.4 | 10.4 | 10.8 |
| 2 | 9.5 | - | 10 | 20 | - | - | - | 0.1 | 1.2 | 0.0 | 9.0 | 8.9 | 0.5 | 10.3 | 10.7 |
| 3 | 9.5 | - | 10 | 35 | - | - | - | 0.1 | 1.2 | 0.0 | 11.1 | 11.1 | 0.0 | 9.4 | 9.5 |
| 4 | 9.5 | - | 40 | 20 | - | - | - | 0.1 | 1.2 | -0.1 | 10.4 | 10.3 | 1.1 | 11.0 | 12.1 |
| 5 | 9.5 | - | 40 | 35 | - | - | - | 0.1 | 1.2 | 0.0 | 8.1 | 8.1 | 0.8 | 8.5 | 9.3 |
| 6 | 9.5 | - | 40 | - | - | - | - | 0.1 | 1.2 | 0.0 | 9.5 | 9.5 | 0.0 | 8.9 | 9.0 |
| 7 | 8 | - | - | - | - | - | - | 0.1 | 1.2 | 0.2 | 4.9 | 5.1 | 0.1 | 5.9 | 6.0 |
| 8 | 8 | - | 10 | 20 | - | - | - | 0.1 | 1.2 | 0.0 | 7.8 | 7.8 | 0.0 | 7.0 | 7.0 |
| 9 | 8 | - | 10 | 35 | - | - | - | 0.1 | 1.2 | -0.1 | 9.6 | 9.4 | 0.0 | 6.7 | 6.7 |
| 10 | 8 | - | 40 | 20 | - | - | - | 0.1 | 1.2 | 0.1 | 6.4 | 6.5 | 0.1 | 7.9 | 8.0 |
| 11 | 8 | - | 40 | 35 | - | - | - | 0.1 | 1.2 | 0.0 | 9.7 | 9.7 | 0.1 | 8.1 | 8.2 |
| 12 | 8 | - | 40 | - | - | - | - | 0.1 | 1.2 | 0.1 | 4.3 | 4.4 | 1.0 | 1.9 | 2.8 |
| 13 | 9.5 | 45 | - | - | - | - | - | 4 | 48.0 | -0.9 | 37.0 | 36.1 | -0.7 | 36.2 | 35.5 |
| 14 | 9.5 | 70 | - | - | - | - | - | 10 | 120.1 | -0.4 | 96.4 | 96.0 | -1.9 | 102.0 | 100.0 |
| 15 | 8 | 45 | - | - | - | - | - | 4 | 48.0 | -0.7 | 29.2 | 28.5 | -0.4 | 35.4 | 35.0 |
| 16 | 8 | 70 | - | - | - | - | - | 10 | 120.1 | -1.5 | 64.1 | 62.6 | -1.5 | 59.3 | 57.8 |
| 17 | 9.5 | - | - | - | 3 | - | - | 0.1 | 1.2 | 0.0 | 12.1 | 12.1 | 1.5 | 9.2 | 10.7 |
| 18 | 9.5 | - | - | - | 3 | 1 | - | 0.1 | 1.2 | 0.2 | 8.6 | 8.8 | 0.8 | 10.3 | 11.0 |
| 19 | 9.5 | - | - | - | - | 1 | - | 0.1 | 1.2 | 0.1 | 7.6 | 7.6 | 0.1 | 8.1 | 8.2 |
| 20 | 9.5 | - | 40 | 35 | - | - | 5 | 0.1 | 6.2 | 6.0 | 16.0 | 21.9 | 6.3 | 13.3 | 19.6 |
| 21 | 9.5 | 70 | - | - | - | - | 5 | 10 | 125.1 | 3.4 | 94.4 | 97.8 | 4.3 | 79.9 | 84.1 |
| 22 | 9.5 | - | - | - | 3 | 1 | 5 | 0.1 | 6.2 | 5.8 | 13.4 | 19.2 | 6.3 | 10.9 | 17.2 |
| 23 | 7 | - | - | - | - | - | - | 0.1 | 1.2 | 0.2 | 2.7 | 2.8 | 0.1 | 3.0 | 3.1 |
| 24 | 7 | 20 | 25 | 25 | 1 | - | - | 0.1 | 1.2 | 0.4 | 0.9 | 1.3 | 0.3 | 1.0 | 1.3 |
| 25 | 8 | 20 | 25 | 25 | 1 | - | - | 0.1 | 1.2 | 0.1 | 8.7 | 8.8 | 0.1 | 8.4 | 8.5 |
| 26 | 9.5 | 20 | 25 | 25 | 1 | - | - | 0.1 | 1.2 | -0.2 | 11.7 | 11.6 | 0.8 | 9.6 | 10.4 |

## Table A.5

## ICP data for in-depth study

| Test # | pH | Ca mg/L | Mg mg/L | Si mg/L | P mg/L | Zn mg/L | NOM mg C/L | $HCO_3^-$ mM | Time (days) | Ca (µg/L) | Mg (µg/L) | Si (µg/L) | P (µg/L) | Zn (µg/L) | Al (µg/L) |
|---|---|---|---|---|---|---|---|---|---|---|---|---|---|---|---|
| 1 | 9.5 | - | - | - | - | - | - | 0.1 | | 0 | 0 | 0 | 0.0 | 0.0 | 0.0 |
| | | | | | | | | | 6 | 26,720 | 215 | 3,350 | 32.6 | 4.8 | 369.4 |
| | | | | | | | | | 14 | 14,500 | 203 | 2,773 | 13.9 | 1.6 | 376.5 |
| | | | | | | | | | 42 | 9,308 | 292 | 5,936 | 11.9 | 1.0 | 290.4 |
| | | | | | | | | | 70 | 4,218 | 144 | 4,345 | 5.3 | 0.3 | 370.6 |
| | | | | | | | | | 102 | 2,595 | 78 | 3,770 | 0.5 | 0.0 | 379.0 |
| | | | | | | | | | 140 | 2,737 | 65 | 3,217 | 0.0 | 0.0 | 383.1 |
| | | | | | | | | | 180 | 2,287 | 55 | 3,323 | 4.8 | 5.1 | 421.9 |
| | | | | | | | | | 232 | 2,201 | 54 | 3,537 | 10.6 | 6.2 | 385.8 |
| 2 | 9.5 | - | 10 | 20 | - | - | - | 0.1 | | 0 | 10,000 | 20,000 | 0.0 | 0.0 | 0.0 |
| | | | | | | | | | 6 | 27,280 | 9,763 | 16,720 | 37.2 | 6.7 | 34.6 |
| | | | | | | | | | 14 | 18,710 | 9,455 | 16,880 | 13.2 | 0.9 | 24.0 |
| | | | | | | | | | 42 | 16,200 | 3,528 | 17,360 | 2.9 | 0.3 | 13.5 |
| | | | | | | | | | 70 | 15,740 | 3,732 | 16,990 | 0.3 | 0.0 | 11.5 |
| | | | | | | | | | 102 | 12,590 | 4,135 | 17,430 | 0.0 | 0.0 | 10.1 |
| | | | | | | | | | 140 | 13,180 | 4,142 | 16,240 | 0.0 | 0.0 | 7.1 |
| | | | | | | | | | 180 | 11,950 | 4,737 | 18,330 | 1.7 | 4.6 | 9.6 |
| | | | | | | | | | 232 | 11,100 | 4,472 | 18,920 | 0.0 | 3.6 | 24.0 |
| 3 | 9.5 | - | 10 | 35 | - | - | - | 0.1 | | 0 | 10,000 | 35,000 | 0.0 | 0.0 | 0.0 |
| | | | | | | | | | 6 | 24,520 | 9,011 | 29,520 | 38.2 | 3.6 | 22.9 |
| | | | | | | | | | 14 | 15,700 | 8,212 | 26,990 | 10.8 | 1.0 | 18.2 |
| | | | | | | | | | 42 | 19,980 | 3,785 | 29,760 | 0.5 | 2.4 | 15.7 |
| | | | | | | | | | 70 | 16,570 | 3,497 | 27,580 | 0.4 | 0.8 | 54.1 |
| | | | | | | | | | 102 | 13,260 | 3,458 | 29,790 | 0.0 | 0.0 | 6.8 |
| | | | | | | | | | 140 | 11,920 | 2,695 | 25,600 | 0.0 | 0.1 | 10.4 |
| | | | | | | | | | 180 | 16,560 | 3,150 | 31,630 | 7.8 | 5.7 | 14.6 |
| | | | | | | | | | 232 | 12,090 | 2,286 | 32,640 | 2.8 | 5.3 | 11.4 |
| 4 | 9.5 | - | 40 | 20 | - | - | - | 0.1 | | 0 | 40,000 | 20,000 | 0.0 | 0.0 | 0.0 |
| | | | | | | | | | 6 | 15,450 | 43,010 | 16,360 | 25.9 | 3.4 | 17.9 |
| | | | | | | | | | 14 | 13,250 | 42,130 | 15,480 | 8.8 | 0.5 | 10.2 |
| | | | | | | | | | 42 | 24,470 | 28,390 | 15,970 | 2.4 | 0.7 | 6.8 |
| | | | | | | | | | 70 | 20,550 | 28,630 | 15,280 | 27.3 | 0.5 | 14.7 |
| | | | | | | | | | 102 | 16,700 | 28,520 | 15,400 | 0.0 | 0.0 | 0.0 |
| | | | | | | | | | 140 | 15,320 | 26,720 | 14,130 | 0.0 | 0.0 | 0.0 |
| | | | | | | | | | 180 | 26,750 | 22,180 | 15,490 | 26.9 | 4.0 | 1.0 |
| | | | | | | | | | 232 | 11,620 | 21,560 | 16,050 | 13.0 | 3.9 | 0.3 |
| 5 | 9.5 | - | 40 | 35 | - | - | - | 0.1 | | 0 | 40,000 | 35,000 | 0.0 | 0.0 | 0.0 |
| | | | | | | | | | 6 | 21,750 | 42,160 | 28,930 | 30.3 | 3.0 | 24.2 |
| | | | | | | | | | 14 | 16,430 | 39,670 | 27,140 | 13.6 | 2.2 | 10.2 |
| | | | | | | | | | 42 | 26,920 | 23,390 | 22,070 | 0.0 | 0.2 | 6.4 |
| | | | | | | | | | 70 | 20,990 | 23,190 | 21,940 | 0.9 | 0.0 | 25.6 |
| | | | | | | | | | 102 | 18,240 | 23,010 | 22,110 | 0.0 | 0.0 | 1.9 |
| | | | | | | | | | 140 | 17,000 | 19,410 | 18,430 | 0.0 | 0.5 | 3.5 |
| | | | | | | | | | 180 | 20,390 | 13,680 | 17,340 | 12.2 | 5.8 | 9.6 |
| | | | | | | | | | 232 | 12,400 | 13,810 | 18,380 | 5.2 | 7.1 | 1.5 |

(continued)

## Table A.5 (continued)

| Test # | pH | Ca mg/L | Mg mg/L | Si mg/L | P mg/L | Zn mg/L | NOM mg C/L | $HCO_3^-$ mM | Time (days) | Ca (μg/L) | Mg (μg/L) | Si (μg/L) | P (μg/L) | Zn (μg/L) | Al (μg/L) |
|---|---|---|---|---|---|---|---|---|---|---|---|---|---|---|---|
| 6 | 9.5 | - | 40 | - | - | - | - | 0.1 | | 0 | 40,000 | 0 | 0.0 | 0.0 | 0.0 |
| | | | | | | | | | 6 | 23,820 | 42,740 | 1,845 | 26.2 | 2.8 | 62.1 |
| | | | | | | | | | 14 | 17,630 | 42,080 | 1,545 | 10.8 | 0.0 | 57.8 |
| | | | | | | | | | 42 | 24,570 | 30,610 | 2,721 | 12.5 | 0.8 | 32.4 |
| | | | | | | | | | 70 | 20,000 | 31,570 | 1,633 | 7.7 | 0.0 | 31.0 |
| | | | | | | | | | 102 | 15,200 | 34,330 | 906 | 0.0 | 0.0 | 15.2 |
| | | | | | | | | | 140 | 13,950 | 29,890 | 611 | 0.0 | 0.0 | 7.9 |
| | | | | | | | | | 180 | 11,670 | 31,000 | 839 | 9.9 | 4.7 | 4.0 |
| | | | | | | | | | 232 | 12,230 | 33,550 | 840 | 6.2 | 3.6 | 7.3 |
| 7 | 8 | - | - | - | - | - | - | 0.1 | | 0 | 0 | 0 | 0.0 | 0.0 | 0.0 |
| | | | | | | | | | 6 | 38,970 | 195 | 3,094 | 18.0 | 5.5 | 321.2 |
| | | | | | | | | | 14 | 20,490 | 222 | 2,336 | 7.2 | 0.2 | 317.8 |
| | | | | | | | | | 42 | 32,150 | 366 | 4,772 | 4.3 | 0.2 | 161.5 |
| | | | | | | | | | 70 | 23,260 | 259 | 3,694 | 0.0 | 0.1 | 224.8 |
| | | | | | | | | | 102 | 19,800 | 168 | 3,233 | 0.0 | 0.0 | 246.6 |
| | | | | | | | | | 140 | 18,580 | 153 | 2,780 | 0.0 | 0.0 | 219.0 |
| | | | | | | | | | 180 | 19,920 | 178 | 3,341 | 8.6 | 4.3 | 289.4 |
| | | | | | | | | | 232 | 25,030 | 184 | 3,692 | 0.0 | 4.1 | 279.3 |
| 8 | 8 | - | 10 | 20 | - | - | - | 0.1 | | 0 | 10,000 | 20,000 | 0.0 | 0.0 | 0.0 |
| | | | | | | | | | 6 | 34,260 | 9,554 | 17,220 | 26.7 | 0.7 | 19.0 |
| | | | | | | | | | 14 | 24,910 | 9,606 | 16,340 | 18.5 | 0.0 | 12.7 |
| | | | | | | | | | 42 | 35,210 | 5,377 | 19,450 | 10.7 | 0.0 | 4.8 |
| | | | | | | | | | 70 | 29,240 | 4,966 | 18,420 | 0.0 | 0.0 | 5.8 |
| | | | | | | | | | 102 | 23,720 | 4,521 | 18,250 | 0.0 | 0.0 | 4.7 |
| | | | | | | | | | 140 | 21,730 | 3,876 | 16,400 | 0.0 | 0.0 | 4.8 |
| | | | | | | | | | 180 | 28,560 | 4,711 | 19,780 | 9.7 | 4.1 | 7.8 |
| | | | | | | | | | 232 | 28,850 | 4,035 | 20,300 | 2.1 | 4.1 | 32.9 |
| 9 | 8 | - | 10 | 35 | - | - | - | 0.1 | | 0 | 10,000 | 35,000 | 0.0 | 0.0 | 0.0 |
| | | | | | | | | | 6 | 35,310 | 10,100 | 33,200 | 33.7 | 4.6 | 8.2 |
| | | | | | | | | | 14 | 27,330 | 9,919 | 29,190 | 24.0 | 1.5 | 8.6 |
| | | | | | | | | | 42 | 33,800 | 5,290 | 29,860 | 9.7 | 0.2 | 1.6 |
| | | | | | | | | | 70 | 28,860 | 4,804 | 29,030 | 0.0 | 0.1 | 5.4 |
| | | | | | | | | | 102 | 23,120 | 4,303 | 30,130 | 0.0 | 0.0 | 1.2 |
| | | | | | | | | | 140 | 21,070 | 3,632 | 28,180 | 0.0 | 0.0 | 1.6 |
| | | | | | | | | | 180 | 32,120 | 4,498 | 33,460 | 5.9 | 6.5 | 4.3 |
| | | | | | | | | | 232 | 33,400 | 4,305 | 35,280 | 3.6 | 4.4 | 2.5 |
| 10 | 8 | - | 40 | 20 | - | - | - | 0.1 | | 0 | 40,000 | 20,000 | 0.0 | 0.0 | 0.0 |
| | | | | | | | | | 6 | 26,580 | 44,290 | 17,530 | 36.7 | 1.4 | 12.0 |
| | | | | | | | | | 14 | 23,160 | 43,960 | 16,430 | 24.8 | 0.0 | 9.6 |
| | | | | | | | | | 42 | 36,550 | 27,430 | 19,120 | 9.7 | 2.8 | 4.6 |
| | | | | | | | | | 70 | 33,550 | 28,460 | 17,780 | 0.0 | 0.0 | 6.8 |
| | | | | | | | | | 102 | 28,580 | 28,250 | 17,550 | 0.0 | 0.0 | 4.1 |
| | | | | | | | | | 140 | 26,680 | 25,820 | 16,120 | 0.0 | 0.0 | 2.9 |
| | | | | | | | | | 180 | 31,450 | 29,960 | 18,800 | 10.7 | 7.2 | 5.5 |
| | | | | | | | | | 232 | 33,360 | 29,410 | 18,730 | 8.9 | 4.6 | 5.0 |

(continued)

## Table A.5 (continued)

| Test # | pH | Ca mg/L | Mg mg/L | Si mg/L | P mg/L | Zn mg/L | NOM mg C/L | $HCO_3^-$ mM | Time (days) | Ca (µg/L) | Mg (µg/L) | Si (µg/L) | P (µg/L) | Zn (µg/L) | Al (µg/L) |
|---|---|---|---|---|---|---|---|---|---|---|---|---|---|---|---|
| 11 | 8 | - | 40 | 35 | - | - | - | 0.1 | | 0 | 40,000 | 35,000 | 0.0 | 0.0 | 0.0 |
| | | | | | | | | | 6 | 27,590 | 43,780 | 30,910 | 33.3 | 3.9 | 14.5 |
| | | | | | | | | | 14 | 24,980 | 41,460 | 29,710 | 19.6 | 2.9 | 10.6 |
| | | | | | | | | | 42 | 35,990 | 26,610 | 29,680 | 8.7 | 2.0 | 2.7 |
| | | | | | | | | | 70 | 32,630 | 26,280 | 28,550 | 0.0 | 0.2 | 1.8 |
| | | | | | | | | | 102 | 25,840 | 26,780 | 29,820 | 0.0 | 0.0 | 0.6 |
| | | | | | | | | | 140 | 24,410 | 23,890 | 27,000 | 0.0 | 0.0 | 0.4 |
| | | | | | | | | | 180 | 34,750 | 29,530 | 32,180 | 13.2 | 3.9 | 1.7 |
| | | | | | | | | | 232 | 36,640 | 29,730 | 33,260 | 5.3 | 3.7 | 1.5 |
| 12 | 8 | - | 40 | - | - | - | - | 0.1 | | 0 | 40,000 | 0 | 0.0 | 0.0 | 0.0 |
| | | | | | | | | | 6 | 29,070 | 42,630 | 2,011 | 24.9 | 1.0 | 92.8 |
| | | | | | | | | | 14 | 23,840 | 43,540 | 1,797 | 11.7 | 0.0 | 115.0 |
| | | | | | | | | | 42 | 39,200 | 30,590 | 3,810 | 6.0 | 0.4 | 58.4 |
| | | | | | | | | | 70 | 35,670 | 31,150 | 2,493 | 0.0 | 1.4 | 204.2 |
| | | | | | | | | | 102 | 29,830 | 31,780 | 1,753 | 0.0 | 0.6 | 116.1 |
| | | | | | | | | | 140 | 33,590 | 27,340 | 1,288 | 0.0 | 0.0 | 110.3 |
| | | | | | | | | | 180 | 35,230 | 31,680 | 1,343 | 17.7 | 4.4 | 113.5 |
| | | | | | | | | | 232 | 31,100 | 32,670 | 1,407 | 6.8 | 5.3 | 72.1 |
| 13 | 9.5 | 45 | - | - | - | - | - | 4 | | 45,000 | 0 | 0 | 0.0 | 0.0 | 0.0 |
| | | | | | | | | | 6 | 3,203 | 148 | 1,482 | 19.1 | 0.0 | 194.7 |
| | | | | | | | | | 14 | 2,655 | 76 | 893 | 13.2 | 0.0 | 141.3 |
| | | | | | | | | | 42 | 1,923 | 65 | 1,494 | 3.0 | 0.0 | 153.6 |
| | | | | | | | | | 70 | 1,790 | 42 | 994 | 0.0 | 0.0 | 92.8 |
| | | | | | | | | | 102 | 1,324 | 23 | 233 | 0.0 | 0.0 | 60.9 |
| | | | | | | | | | 140 | 1,493 | 17 | 169 | 0.0 | 0.0 | 133.0 |
| | | | | | | | | | 180 | 1,333 | 30 | 281 | 9.1 | 6.6 | 42.4 |
| | | | | | | | | | 232 | 1,468 | 32 | 255 | 2.5 | 3.2 | 42.6 |
| 14 | 9.5 | 70 | - | - | - | - | - | 10 | | 70,000 | 0 | 0 | 0.0 | 0.0 | 0.0 |
| | | | | | | | | | 6 | 1,338 | 132 | 1,474 | 32.8 | 0.0 | 178.9 |
| | | | | | | | | | 14 | 1,013 | 58 | 804 | 7.2 | 0.0 | 151.7 |
| | | | | | | | | | 42 | 840 | 55 | 1,542 | 16.3 | 0.0 | 204.8 |
| | | | | | | | | | 70 | 838 | 38 | 1,748 | 0.0 | 6.1 | 658.1 |
| | | | | | | | | | 102 | 706 | 20 | 386 | 0.0 | 0.0 | 101.6 |
| | | | | | | | | | 140 | 1,096 | 23 | 938 | 0.0 | 0.0 | 69.6 |
| | | | | | | | | | 180 | 1,013 | 36 | 397 | 8.8 | 3.8 | 69.4 |
| | | | | | | | | | 232 | 962 | 32 | 398 | 9.8 | 2.9 | 70.0 |
| 15 | 8 | 45 | - | - | - | - | - | 4 | | 45,000 | 0 | 0 | 0.0 | 0.0 | 0.0 |
| | | | | | | | | | 6 | 49,320 | 163 | 904 | 22.0 | 2.9 | 119.3 |
| | | | | | | | | | 14 | 53,220 | 122 | 761 | 9.1 | 2.0 | 116.6 |
| | | | | | | | | | 42 | 36,220 | 97 | 1,204 | 7.4 | 0.1 | 191.9 |
| | | | | | | | | | 70 | 19,720 | 69 | 1,011 | 0.0 | 0.0 | 148.7 |
| | | | | | | | | | 102 | 16,340 | 47 | 295 | 0.0 | 0.0 | 104.5 |
| | | | | | | | | | 140 | 12,400 | 38 | 154 | 0.0 | 0.0 | 82.2 |
| | | | | | | | | | 180 | 9,693 | 69 | 398 | 14.4 | 4.8 | 80.2 |
| | | | | | | | | | 232 | 11,160 | 51 | 283 | 6.3 | 3.6 | 66.2 |

(continued)

## Table A.5 (continued)

| Test # | pH | Ca mg/L | Mg mg/L | Si mg/L | P mg/L | Zn mg/L | NOM mg C/L | $HCO_3^-$ mM | Time (days) | Ca (μg/L) | Mg (μg/L) | Si (μg/L) | P (μg/L) | Zn (μg/L) | Al (μg/L) |
|---|---|---|---|---|---|---|---|---|---|---|---|---|---|---|---|
| 16 | 8 | 70 | - | - | - | - | - | 10 | | 70,000 | 0 | 0 | 0.0 | 0.0 | 0.0 |
| | | | | | | | | | 6 | 26,690 | 160 | 775 | 15.1 | 0.0 | 51.0 |
| | | | | | | | | | 14 | 20,370 | 56 | 389 | 5.6 | 0.0 | 42.6 |
| | | | | | | | | | 42 | 6,391 | 69 | 1,024 | 8.5 | 0.0 | 158.0 |
| | | | | | | | | | 70 | 4,014 | 59 | 1,402 | 0.0 | 0.7 | 140.5 |
| | | | | | | | | | 102 | 4,052 | 55 | 302 | 0.0 | 0.0 | 88.2 |
| | | | | | | | | | 140 | 3,595 | 33 | 238 | 0.0 | 0.0 | 67.0 |
| | | | | | | | | | 180 | 6,092 | 44 | 339 | 15.5 | 3.9 | 63.0 |
| | | | | | | | | | 232 | 5,938 | 47 | 278 | 4.8 | 3.3 | 58.7 |
| 17 | 9.5 | - | - | - | 3 | - | - | 0.1 | | 0 | 0 | 0 | 3,000.0 | 0.0 | 0.0 |
| | | | | | | | | | 6 | 12,470 | 105 | 1,291 | 2,784.0 | 0.0 | 218.0 |
| | | | | | | | | | 14 | 9,060 | 89 | 1,239 | 3,020.0 | 0.0 | 146.3 |
| | | | | | | | | | 42 | 17,550 | 254 | 5,043 | 232.0 | 0.0 | 204.2 |
| | | | | | | | | | 70 | 11,480 | 169 | 3,841 | 509.1 | 2.4 | 383.1 |
| | | | | | | | | | 102 | 6,178 | 58 | 1,635 | 1,323.0 | 0.0 | 79.2 |
| | | | | | | | | | 140 | 4,603 | 28 | 700 | 1,916.0 | 0.0 | 48.4 |
| | | | | | | | | | 180 | 18,070 | 262 | 5,027 | 200.3 | 8.0 | 289.3 |
| | | | | | | | | | 232 | 15,050 | 223 | 5,306 | 209.0 | 3.7 | 229.7 |
| 18 | 9.5 | - | - | - | 3 | 1 | - | 0.1 | | 0 | 0 | 0 | 3,000.0 | 1,000.0 | 0.0 |
| | | | | | | | | | 6 | 12,220 | 85 | 1,153 | 2,261.0 | 424.8 | 210.7 |
| | | | | | | | | | 14 | 7,259 | 77 | 901 | 2,958.0 | 1,069.0 | 114.1 |
| | | | | | | | | | 42 | 13,850 | 160 | 2,849 | 1,595.0 | 605.3 | 173.0 |
| | | | | | | | | | 70 | 12,500 | 168 | 3,160 | 1,568.0 | 556.2 | 262.8 |
| | | | | | | | | | 102 | 11,150 | 150 | 3,287 | 1,471.0 | 41.7 | 146.7 |
| | | | | | | | | | 140 | 11,070 | 147 | 3,317 | 1,328.0 | 33.9 | 157.0 |
| | | | | | | | | | 180 | 17,780 | 185 | 5,128 | 720.4 | 231.2 | 329.5 |
| | | | | | | | | | 232 | 18,470 | 218 | 4,840 | 733.4 | 93.7 | 226.1 |
| 19 | 9.5 | - | - | - | - | 1 | - | 0.1 | | 0 | 0 | 0 | 0.0 | 1,000.0 | 0.0 |
| | | | | | | | | | 6 | 33,450 | 161 | 3,323 | 21.4 | 356.5 | 366.6 |
| | | | | | | | | | 14 | 19,080 | 139 | 2,466 | 5.3 | 359.5 | 326.5 |
| | | | | | | | | | 42 | 18,740 | 256 | 5,225 | 9.1 | 63.7 | 305.0 |
| | | | | | | | | | 70 | 13,660 | 218 | 3,921 | 0.0 | 93.8 | 370.3 |
| | | | | | | | | | 102 | 8,303 | 139 | 3,256 | 0.0 | 51.7 | 234.6 |
| | | | | | | | | | 140 | 7,724 | 109 | 2,604 | 0.0 | 24.0 | 184.7 |
| | | | | | | | | | 180 | 7,479 | 99 | 2,837 | 6.8 | 21.7 | 153.1 |
| | | | | | | | | | 232 | 4,995 | 85 | 2,931 | 2.4 | 9.2 | 124.7 |
| 20 | 9.5 | - | 40 | 35 | - | - | 5 | 0.1 | | 0 | 40,000 | 35,000 | 0.0 | 0.0 | 0.0 |
| | | | | | | | | | 6 | 23,310 | 42,530 | 29,000 | 156.9 | 16.7 | 68.1 |
| | | | | | | | | | 14 | 14,230 | 40,700 | 27,850 | 144.3 | 18.8 | 52.4 |
| | | | | | | | | | 42 | 26,560 | 25,220 | 26,150 | 95.6 | 13.9 | 35.7 |
| | | | | | | | | | 70 | 22,460 | 25,170 | 26,670 | 91.4 | 13.3 | 33.9 |
| | | | | | | | | | 102 | 18,730 | 26,570 | 29,610 | 92.6 | 10.3 | 30.8 |
| | | | | | | | | | 140 | 16,610 | 22,440 | 24,690 | 66.8 | 6.7 | 16.6 |
| | | | | | | | | | 180 | 24,110 | 27,450 | 30,150 | 107.5 | 13.1 | 28.1 |
| | | | | | | | | | 232 | 27,080 | 28,200 | 31,610 | 79.6 | 13.8 | 28.2 |

(continued)

## Table A.5 (continued)

| Test # | pH | Ca mg/L | Mg mg/L | Si mg/L | P mg/L | Zn mg/L | NOM mg C/L | $HCO_3^-$ mM | Time (days) | Ca (μg/L) | Mg (μg/L) | Si (μg/L) | P (μg/L) | Zn (μg/L) | Al (μg/L) |
|---|---|---|---|---|---|---|---|---|---|---|---|---|---|---|---|
| 21 | 9.5 | 70 | - | - | - | - | 5 | 10 | | 70,000 | 0 | 0 | 0.0 | 0.0 | 0.0 |
| | | | | | | | | | 6 | 5,492 | 258 | 1,773 | 93.0 | 0.3 | 185.4 |
| | | | | | | | | | 14 | 3,160 | 170 | 1,253 | 55.4 | 0.0 | 162.9 |
| | | | | | | | | | 42 | 2,429 | 144 | 2,376 | 52.4 | 0.0 | 265.5 |
| | | | | | | | | | 70 | 2,149 | 122 | 2,837 | 35.6 | 0.0 | 294.5 |
| | | | | | | | | | 102 | 2,090 | 101 | 1,515 | 37.5 | 0.0 | 208.2 |
| | | | | | | | | | 140 | 2,132 | 87 | 1,456 | 27.6 | 0.0 | 197.1 |
| | | | | | | | | | 180 | 2,464 | 89 | 1,442 | 35.3 | 4.2 | 197.4 |
| | | | | | | | | | 232 | 2,369 | 97 | 1,324 | 22.1 | 3.3 | 201.2 |
| 22 | 9.5 | - | - | - | 3 | 1 | 5 | 0.1 | | 0 | 0 | 0 | 3,000.0 | 1,000.0 | 0.0 |
| | | | | | | | | | 6 | 18,110 | 135 | 1,627 | 2,447.0 | 1,049.0 | 249.4 |
| | | | | | | | | | 14 | 13,840 | 150 | 1,649 | 2,417.0 | 1,147.0 | 156.6 |
| | | | | | | | | | 42 | 21,570 | 236 | 3,433 | 1,933.0 | 820.3 | 154.2 |
| | | | | | | | | | 70 | 18,220 | 222 | 3,550 | 2,409.0 | 823.9 | 154.3 |
| | | | | | | | | | 102 | 16,800 | 165 | 3,508 | 2,328.0 | 791.4 | 155.7 |
| | | | | | | | | | 140 | 16,010 | 172 | 3,479 | 2,394.0 | 672.3 | 165.0 |
| | | | | | | | | | 180 | 15,820 | 152 | 3,988 | 2,550.0 | 646.1 | 241.0 |
| | | | | | | | | | 232 | 16,060 | 151 | 4,304 | 2,453.0 | 586.4 | 265.8 |
| 23 | 7 | - | - | - | - | - | - | 0.1 | | 0 | 0 | 0 | 0.0 | 0.0 | 0.0 |
| | | | | | | | | | 6 | 25,790 | 376 | 3,339 | 22.3 | 3.4 | 163.0 |
| | | | | | | | | | 14 | 22,450 | 293 | 2,749 | 12.5 | 0.0 | 166.1 |
| | | | | | | | | | 42 | 37,160 | 412 | 5,662 | 8.2 | 0.6 | 65.5 |
| | | | | | | | | | 70 | 34,370 | 309 | 4,746 | 0.0 | 1.3 | 167.7 |
| | | | | | | | | | 102 | 26,790 | 237 | 4,650 | 0.0 | 0.0 | 107.4 |
| | | | | | | | | | 140 | 25,090 | 226 | 4,328 | 0.0 | 0.0 | 109.4 |
| | | | | | | | | | 180 | 29,300 | 246 | 5,061 | 16.9 | 5.1 | 78.0 |
| | | | | | | | | | 232 | 26,510 | 222 | 5,144 | 2.0 | 3.9 | 96.1 |
| 24 | 7 | 20 | 25 | 25 | 1 | - | - | 0.1 | | 20,000 | 25,000 | 25,000 | 1,000.0 | 0.0 | 0.0 |
| | | | | | | | | | 6 | 45,360 | 29,560 | 22,330 | 1,074.0 | 2.5 | 17.1 |
| | | | | | | | | | 14 | 38,930 | 28,320 | 21,010 | 1,088.0 | 0.3 | 13.8 |
| | | | | | | | | | 42 | 58,850 | 17,810 | 21,940 | 708.2 | 0.3 | 2.6 |
| | | | | | | | | | 70 | 50,110 | 16,670 | 21,710 | 636.6 | 0.4 | 2.4 |
| | | | | | | | | | 102 | 40,970 | 15,810 | 22,390 | 638.1 | 4.3 | 2.5 |
| | | | | | | | | | 140 | 38,370 | 14,280 | 20,590 | 615.9 | 0.0 | 2.4 |
| | | | | | | | | | 180 | 54,160 | 17,130 | 24,620 | 812.1 | 4.8 | 4.4 |
| | | | | | | | | | 232 | 58,300 | 17,160 | 25,300 | 760.1 | 3.5 | 3.6 |
| 25 | 8 | 20 | 25 | 25 | 1 | - | - | 0.1 | | 20,000 | 25,000 | 25,000 | 1,000.0 | 0.0 | 0.0 |
| | | | | | | | | | 6 | 42,040 | 27,710 | 21,420 | 1,051.0 | 3.8 | 10.4 |
| | | | | | | | | | 14 | 35,140 | 25,430 | 20,460 | 1,031.0 | 0.6 | 9.2 |
| | | | | | | | | | 42 | 36,450 | 16,400 | 38,930 | 754.4 | 0.1 | 0.8 |
| | | | | | | | | | 70 | 37,950 | 17,100 | 23,370 | 749.4 | 0.5 | 51.6 |
| | | | | | | | | | 102 | 33,070 | 16,920 | 23,130 | 717.7 | 0.0 | 1.4 |
| | | | | | | | | | 140 | 31,070 | 15,030 | 20,760 | 870.1 | 0.0 | 1.4 |
| | | | | | | | | | 180 | 43,760 | 17,460 | 25,210 | 892.5 | 4.3 | 3.5 |
| | | | | | | | | | 232 | 47,910 | 18,240 | 26,240 | 832.1 | 3.4 | 3.4 |
| 26 | 9.5 | 20 | 25 | 25 | 1 | - | - | 0.1 | | 20,000 | 25,000 | 25,000 | 1,000.0 | 0.0 | 0.0 |
| | | | | | | | | | 6 | 34,100 | 29,010 | 22,170 | 1,082.0 | 1.2 | 21.3 |
| | | | | | | | | | 14 | 29,590 | 27,150 | 20,770 | 913.0 | 0.0 | 15.6 |
| | | | | | | | | | 42 | 31,920 | 17,560 | 20,890 | 723.9 | 0.0 | 5.8 |
| | | | | | | | | | 70 | 28,370 | 18,820 | 22,020 | 751.9 | 0.0 | 1.6 |
| | | | | | | | | | 102 | 26,250 | 18,640 | 22,420 | 753.0 | 0.0 | 1.2 |
| | | | | | | | | | 140 | 24,940 | 16,460 | 20,710 | 747.4 | 0.0 | 0.0 |
| | | | | | | | | | 180 | 47,510 | 14,910 | 21,620 | 334.3 | 3.8 | 1.7 |
| | | | | | | | | | 232 | 53,020 | 17,760 | 25,380 | 739.0 | 3.7 | 1.4 |

# APPENDIX B: DIGITAL IMAGES OF SPECIMENS AT END OF IN-DEPTH STUDY

**Figure B.1: Digital Images of Specimens from Test #4 (pH 9.5, 40 mg/L Mg, 20 mg/L Si, 0.1 mM $HCO_3^-$).** Specimen '4b' has 'glue' oozing from simulated crack.

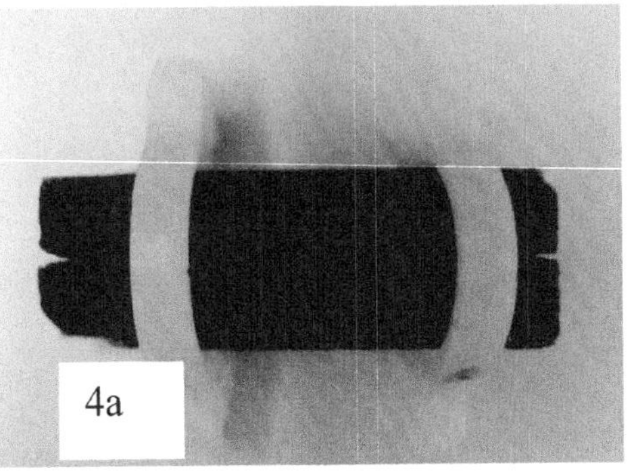


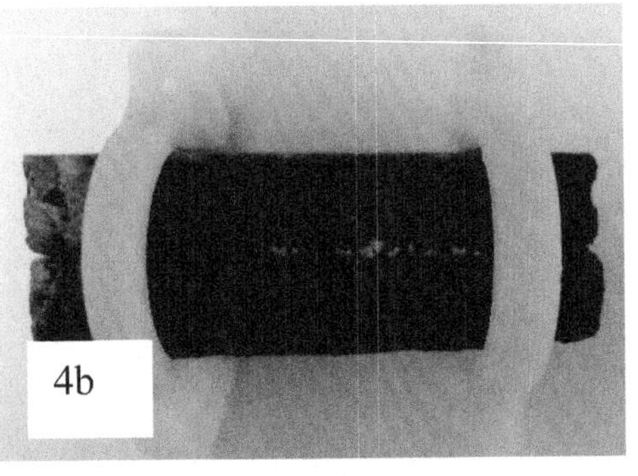


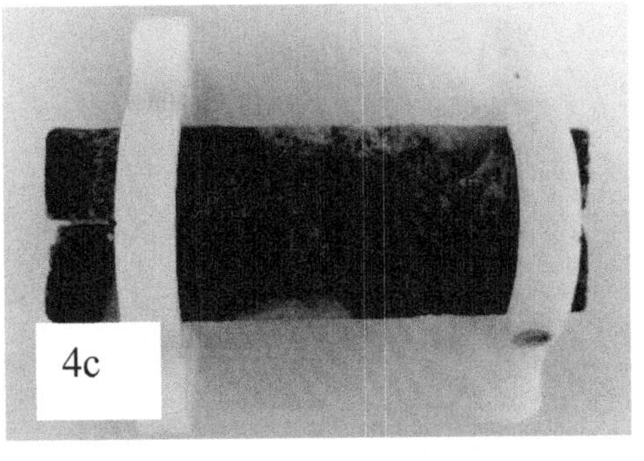

**Figure B.2: Digital Images of Specimens from Test #5 (pH 9.5, 40 mg/L Mg, 35 mg/L Si, 0.1 mM $HCO_3^-$).** 'Glue' can be seen protruding from simulated crack in specimen '5a' and '5b'.

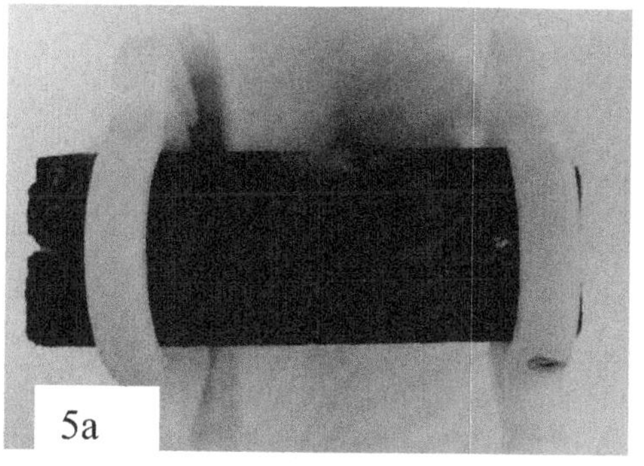


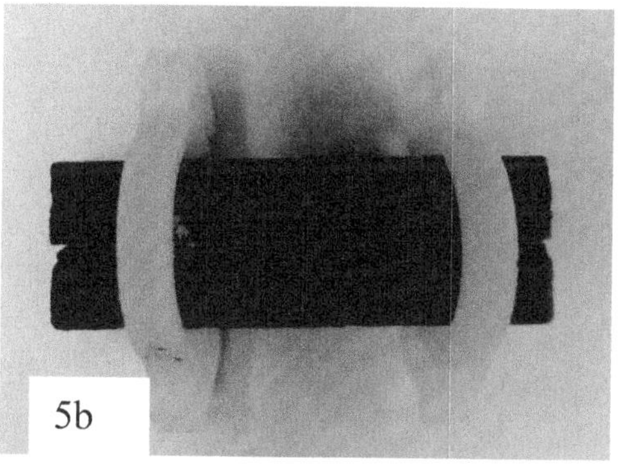


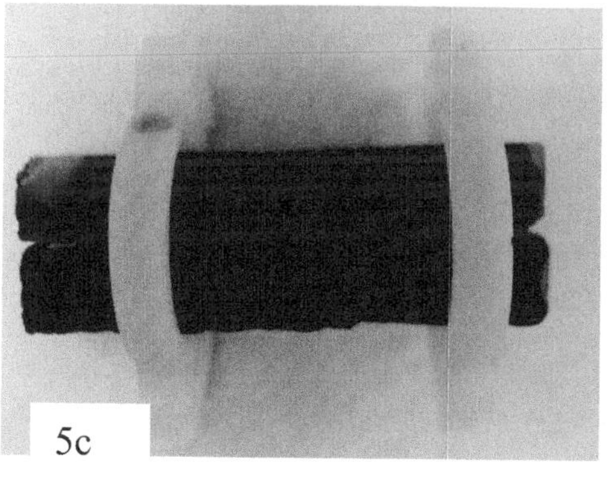

**Figure B.3: Digital Images of Specimens from Test #18 (pH 9.5, 3 mg/L P, 1 mg/L Zn, 0.1 mM $HCO_3^-$).** White precipitate can be seen on the surface of specimen '18b' and in the simulated crack of specimen '18c'.

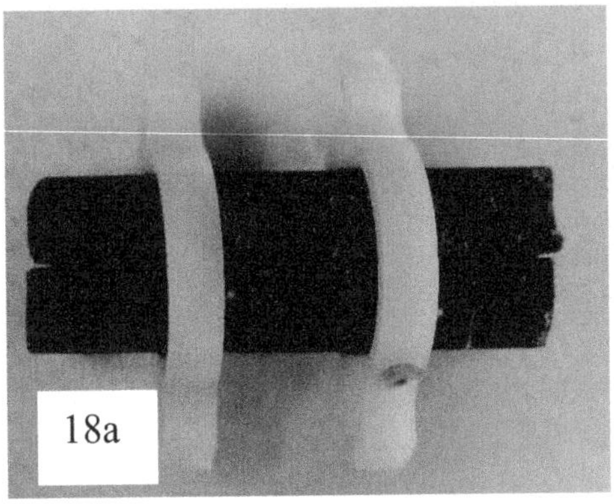


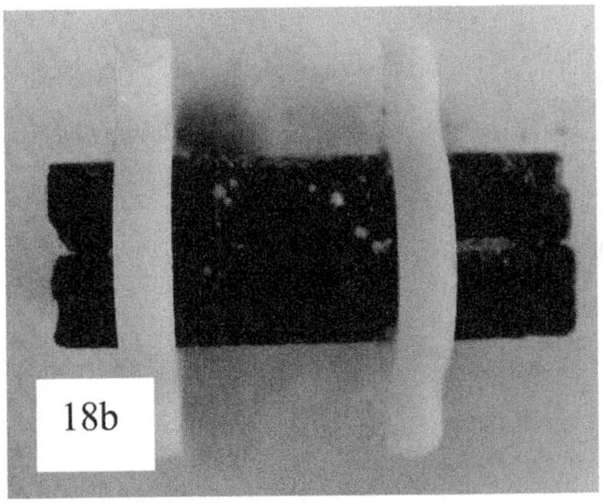


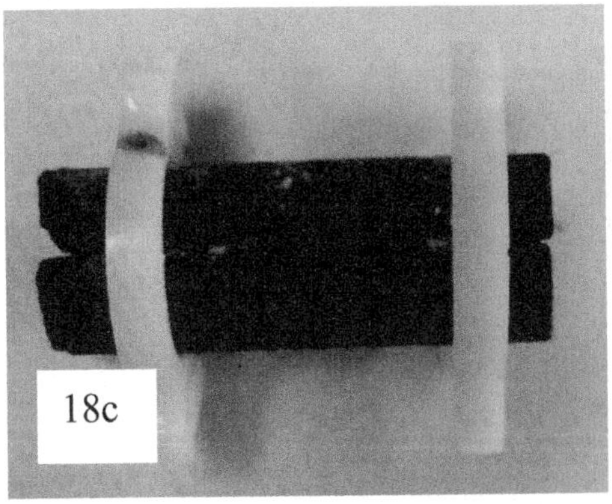

# APPENDIX C: DIGITAL IMAGES FROM CARBONATION TESTING

**Figure C.1: Carbonation Comparison of Specimens from Test #9 (pH 8, 10 mg/L Mg, 35 mg/L Si, 0.1 mM $HCO_3^-$).** Pink color indicates neither specimen is carbonated.

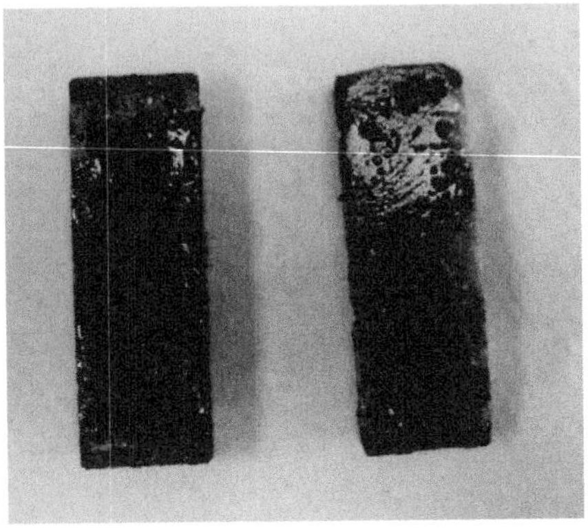

9b

9c

**Figure C.2: Image of Specimen from Test #26 (pH 9.5, 25 mg/L Mg, 25 mg/L Si, 20 mg/L Ca, 1 mg/L P, 0.1 mM $HCO_3^-$).** Lack of pink color indicates specimen is completely carbonated.

**Figure C.3: Image of Specimen from Test #3 (pH 9.5, 10 mg/L Mg, 35 mg/L Si, 0.1 mM $HCO_3^-$).**
Pink color indicates specimen is not carbonated.

# REFERENCES


Aldea, C-M., Song, W-J., Popovics, J.S., and S. P. Shah. 2000. Extent of Healing of Cracked Normal Strength Concrete. Journal of Materials in Civil Engineering. Feb: 92-96.

APHA, AWWA, and WEF (American Public Health Association, American Water Works Association, and Water Environment Federation). 1998. *Standard Methods for theExamination of Water and Wastewater*, 20th ed. Washington, DC: APHA, AWWA, and WEF.

AWWA. 2001. Dawn of the Replacement Era: Reinvesting in Drinking Water Infrastructure. AWWA Government Affairs Office, Washington DC.

AwwaRF and DVGW-TZW. 1996. *Internal Corrosion of Water Distribution Systems*. AwwaRF, Denver, CO.

Berend, K. and T. Trouwborst. 1999. Cement-Mortar Pipes as a Source of Aluminum. Journal American Water Works Association, 91(7), 91-100.

Brindley, G.W., Bish, D.L., and Wan, Hsien-Ming. 1977. The Nature of Kerolite, its Relation to Talc and Stevensite. Mineralogical Magazine 41: 443-452.

Broomfield, J.P. 2000. Carbonation and its Effects in Reinforced Concrete. Materials Performance, 39(1): 64-67.

California Institute of Technology. 2005. Raman Spectroscopy Database (Online). Available: <http://minerals.gps.caltech.edu/files/raman/Caltech_data/index.htm> [cited September 2006]

Christ, C.L., Hostetler, P.B., and Siebert, R.M. 1973. Studies in the System $MgO-SiO_2-CO_2-H_2O$ (III): The Activity-Product Constant of Sepiolite. American Journal of Science 273: 65-83.

Clear, C.A. 1985. The Effects of Autogenous Healing Upon Leakage of Water Through Cracks in Concrete. Cem. And Concr. Ass., Tech Rep. 559 May.

Deng, C. S., Breen, C., Yarwood, J., Habesch, S., Phipps, J., Craster, B. and Maitland, G. 2002. Ageing of Oilfield Cement at High Humidity: A Combined FEG-ESEM and Raman Microscopic Investigation. Journal of Materials Chemistry 12(10): 3105-3112.

Dingus, M., Haven, J., and R. Austin. 2002. Nondestructive, Noninvasive Assessment of Underground Pipelines. AwwaRF Final Report. AWWA. Denver.

Douglas, B. D., Merrill, D. T. and Catlin, J. O. 1996. Water Quality Deterioration From Corrosion of Cement-Mortar Linings. Journal American Water Works Association 88(7): 99-107.

Edvarsen, C. 1999. Water Permeability and Autogenous Healing of Cracks in Concrete. ACI Materials Journal, 96 (4): 448-454.

Gunnlaugsson, E., and Einarsson, A. 1989. Magnesium-Silicate Scaling in Mixture of Geothermal Water and Deaerated Fresh Water in a District Heating System. Geothermics, 18(1/2): 113-120.

Hearn, N. 1998. Self-sealing, Autogenous Healing and Continued Hydration: What is the Difference? Materials and Structures 31 (212) 563-567.

Hearn, N., and Morley, C.T. 1997. Self-sealing Property of Concrete – Experimental Evidence. Materials and Structures 30 (201): 404-411.

Jackson, R.Z., Pitt, C., and R. Scabo. 1992. Nondestructive Testing of Water Mains Physical Integrity. AwwaRF Final Report. AWWA. Denver.

Kristiansen, H. 1977. Corrosion of Asbestos-Cement Pipes. Vatten, 33(1), 53-61.

Kristmannsdottir, H., Olafsson, M., and Thorhallsson, S. 1989. Magnesium-Silicate Scaling in District Heating Systems in Iceland. Geothermics, 18(1/2): 191-198.

Kvech, S. J., and Edwards, M. 2002. The Role of Aluminosilicate Deposits in Lead and Copper Corrosion. Journal American Water Works Association, 93(11), 104-112.

McNeill, L.S., and M. Edwards. 2003. Degradation of Drinking Water Plant Infrastructure from Enhanced Coagulation. Journal of Infrastructure Systems. 9 (4):145-156.

Munday, J.G.L., Sangha, C.M., and Dhir, R.K. 1974. Comparative Study of Autogenous Healing of Different Concretes. First Australian Conference on Engineering Materials. Syndey, Australia. Pp. 177-189.

Neville, A. 2002. Autogenous Healing-A Concrete Miracle? Concrete International, 24 (11).

Parks, J.L. and Edwards, M. 2007. Boron Removal via Formation of Magnesium Silicate Solids During Precipitative Softening. Journal of Environmental Engineering. Forthcoming.

Price, M.L., Snoeyink, V.L., Clark, S.C., Kernkamp, A.L., and Musgrove, J.L. 1997. Control of Magnesium Silicate Scaling in Austin's Lime Softened Water. Proceedings 1997 WQTC, Nov, 1997, Denver, CO.

Ramm, W., and M. Biscoping. 1998. Autogenous Healing and Reinforcement Corrosion of Water-Penetrated Separation Cracks in Reinforced Concrete. Nuclear Engineering and Design, 179, 191-200.

Russell, L.L. 1976. Chemical Aspects of Groundwater Recharge with Wastewaters. Ph.D. Thesis, University of California, Berkeley.

Santhanam, M., Cohen, M.D., and Olek, J. 2002. Mechanism of Sulphate Attack: A Fresh Look. Pt. 1. Summary of Experimental Results. Cement and Concrete Research, 32(6): 915-921.

Schock, M.R. and R.W. Buelow. 1981. The Behavior of Asbestos-Cement Pipe Under Various Water Conditions. 2. Theoretical Considerations. Journal American Water Works Association, 73(12), 636-651.

Slats 2002. Personal communication.

Universita di Siena. 2005. Raman Spectroscopy Database (Online). Available: <http://www.dst.unisi.it/geofluids/raman/spectrum_frame.htm>. [cited September 2006]

U.S. Department of Transportation. 2002. Corrosion Cost and Preventive Strategies in the United States. No. FHWA-RD-01-156. McLean, VA: Federal Highway Administration.

Wagner, E.F. (1974) Autogenous Healing of Cracks in Cement-Mortar Linings for Gray-Iron and Ductile Iron Water Pipe. Journal American Water Works Association 66(6): 358-360.

# ABBREVIATIONS

| | |
|---|---|
| APHA | American Public Health Association |
| AWWA | American Water Works Association |
| AwwaRF | Awwa Research Foundation |
| $\gamma_{HCO_3^-}$ | activity coefficient of bicarbonate |
| $\gamma_{Ca^{2+}}$ | activity coefficient of calcium |
| °C | degrees Celcius |
| d | day |
| °F | degrees Fahrenheit |
| FE-SEM | field emission – scanning electron microscope |
| IAP | ion activity product |
| ICP-ES | inductively coupled plasma - emission spectrometry |
| ICP-MS | inductively coupled plasma - mass spectrometry |
| in | inch |
| L | liter |
| LI | Langelier Index |
| m | meter |
| M | molar |
| mg | milligram |
| mg/L | milligram per liter |
| mL | milliliter |
| mm | millimeter |
| mM | millimolar |
| mo | month |

| | |
|---|---|
| µmho | micro-mho |
| | |
| N | Newton |
| NOM | natural organic matter |
| | |
| PAC | Project Advisory Committee |
| $pH_a$ | measured bulk pH |
| $pH_s$ | pH at which the solution is saturated with $CaCO_{3(s)}$ |
| $pK_s$ | solubility constant for calcite at a given temperature. |
| pp. | pages |
| ppb | parts per billion |
| ppm | parts per million |
| PVC | polyvinyl chloride |
| | |
| RH | relative humidity |
| rpm | revolutions per minute |
| | |
| s | second |
| SEM-EDS | scanning electron microscopy with energy dispersive spectroscopy |
| SI | saturation index |
| SMCL | secondary maximum contaminant level |
| SSE | self sealing effect |
| | |
| TOC | total organic carbon |
| | |
| wk | week |
| WQTC | Water Quality Technology Conference |
| | |
| yr | year |